Experimental Techniques in
Bacterial Genetics

The Jones and Bartlett Series in Biology

Experimental Techniques in Bacterial Genetics

Stanley R. Maloy

Department of Microbiology
University of Illinois, Urbana

JONES AND BARTLETT PUBLISHERS

BOSTON

Editorial, Sales, and Customer Service Offices

Jones and Bartlett Publishers
20 Park Plaza
Boston, MA 02116

Printed in the United States of America
10 9 8 7 6 5 4 3 2 1

Library of Congress Cataloging-in-Publication Data

Maloy, Stanley.
Experimental techniques in bacterial genetics / Stanley Maloy.
   p. cm.
Includes bibliographies and index.
ISBN 0-86720-118-5
1. Bacterial genetics--Experiments.   I. Title.
QH434.M35 1989
589.9'015072–dc20   89-11078
                                                        CIP

Manuscript editor Leesa Stanion
Text and cover designs Rafael Millán
Production Marina Publications
Cover illustration E. coli © 1983 Designer Genes Posters, Ltd.
All rights reserved. Available on T-shirts, posters and postcards from Carolina Biological Supply Co., 2700 York Rd., Burlington, NC 27215 (1-800-334-5551).

Preface

The experiments in this manual use both *in vivo* genetic techniques and *in vitro* molecular techniques to study gene structure, function, and regulation in bacteria. Both approaches are useful for studying bacterial genetics: the function and regulation of a gene can be directly determined *in vivo*, and the detailed molecular structure of a gene can be determined *in vitro*. Both *in vivo* and *in vitro* techniques are also useful when cloning foreign genes in bacteria: in order to clone a foreign gene in bacteria it is sometimes necessary to genetically manipulate the host to stabilize the clone or to increase expression of the cloned gene (e.g., to eliminate host recombination functions, proteases, or other enzymes). Although the experiments in this manual use *S. typhimurium* and *E. coli*, many of the methods can be used for studying the molecular genetics of other organisms as well.

This manual is based on a course I have taught for four years at the University of Illinois, Urbana. The course is designed to give upper division undergraduates and graduate students hands-on experience with modern techniques in bacterial genetics and molecular biology. An enormous number of variations of molecular techniques are available. Every molecular geneticist likes some techniques and dislikes others, but no two molecular geneticists seem to have the same preferences. "New and improved" techniques are constantly being developed, but most of these techniques rely on a few basic skills and principles. The methods in this lab manual are not meant to demonstrate every possible technique, but to provide experience with a variety of basic genetic and molecular techniques. With this background a student should be able to quickly learn new techniques and develop personal likes and dislikes.

In addition to teaching experimental techniques, another goal of this course is to teach students how to design and interpret such experiments. The strategy for teaching these skills is to require students to think about their experiments instead of just following instructions. Students construct operon fusions with *S. typhimurium* genes, then use genetic and molecular techniques to characterize the regulation of the mutated gene and the nature of the gene products. Although each experimental technique is described in the lab manual, the students isolate different mutants, so they need to plan the specific experimental details required for their mutants. If an experiment does not work (as often happens in research labs), I consult with the students to try to figure out what went wrong and encourage them to try again. In order to interpret their results, I encourage students to look up references in the literature and compare their results with published results.

This format provides an intensive, individualized lab experience. Unlike many typical lab courses, it will not be possible to do everything during regular laboratory hours. Usually students will only have to drop by the lab for a few minutes on days the lab does not meet to check their plates or to stop an agarose gel, etc., but a few experiments involve more lengthy procedures on odd-days which may require special arrangements.

In order to understand how an experiment works, it is important to know what reagents were used and why. The specific reagents required for each experiment are listed at the end of each experiment. In addition, for the convenience of the instructor all of the reagents are also listed alphabetically in the Appendix. I have mentioned specific brand names in some experiments because I have had satisfying experiences with those products -- it is not meant to imply that the generic equivalent will not work equally well.

Several of the experiments in this lab manual use radioisotopes. Some schools do not presently use radioisotopes in teaching laboratories. However, radioisotopes are used extensively in molecular biology, so I think use of radioisotopes is essential for a realistic laboratory experience. Relatively low levels of radioactivity are used in these experiments and only ^{35}S is used, an isotope with sufficiently low energy that shields are not required but high enough energy that it can be directly monitored with a thin-window Geiger counter. Nevertheless, most of the experiments in this manual do not use radioactivity. These experiments could be easily used for a "nuclear-free" laboratory including many techniques in bacterial genetics, cloning and restriction analysis.

This manual assumes that the students have some previous theoretical background in genetics and biochemistry. Short explanations of important concepts and the rationale for each experiment are included, but this lab manual is not meant to provide an extensive theoretical background in bacterial genetics or molecular biology. More details on the theory can be obtained from the references listed or from Microbiology, Genetics, Molecular Biology, or Biochemistry textbooks.

STANLEY R. MALOY
Department of Microbiology
University of Illinois, Urbana

Acknowledgments

All of the experiments done in this course are modifications of published protocols that are used in my lab. I learned many of these procedures in the Advanced Bacterial Genetics course taught at Cold Spring Harbor Laboratory by Tom Silhavy, Lynn Enquist, and Mike Berman, and while working in John Roth's lab. For many of these procedures the following excellent sources were my initial references:

Davis, R., D. Botstein, and J. Roth. 1980. *Advanced Bacterial Genetics*. Cold Spring Harbor Laboratory, NY.

Maniatis, T., E. Fritsch, and J. Sambrook. 1982. *Molecular Cloning*. Cold Spring Harbor Laboratory, NY.

Roth, J. 1970. Genetic techniques in studies of bacterial metabolism. *Methods Enzymol.* 17: 1-35.

Silhavy, T., L. Enquist, and M. Berman. 1984. *Experiments with Gene Fusions*. Cold Spring Harbor Laboratory, NY.

Several reviewers made excellent suggestions, including Jon Beckwith (Harvard Medical School), Andrew Kropinski (Queen's University, Ontario), Paul Matsudaira (MIT), Charles Miller (Case Western Reserve University), and Valley Stewart (Cornell University). In addition, many useful suggestions were made by the previous teaching assistants for this course and my graduate students: Li-Mei Chen, Greg Deno, Don Hahn, Craig Kent, Eunhee Lee, Min-Ken Liao, Bob Lloyd, Rik Myers, Nick Santaros, Paula Spicer, and Ming Te Yang. Michele Beaudet and Joe Pogliano carefully proofread the final version of this manual.

Table of Contents

COURSE OUTLINE

A suggested outline for a lab course using this manual is shown below:

EXPERIMENTS

a. Isolate a Mud insertion mutant in *S. typhimurium* and determine its metabolic effect (e.g., the auxotrophic requirement).
b. Map the insertion mutation on the *S. typhimurium* chromosome.
c. Isolate regulatory mutants that affect expression of the Mud insertion mutant.
d. Construct a physical map of the chromosomal insertion mutation by Southern blotting.
e. Isolate a clone that complements the insertion mutation and characterize the cloned gene by mutagenesis, restriction mapping, and expression of the plasmid encoded proteins.
f. Subclone a fragment of the complementing gene into M13 and determine the DNA sequence of the cloned fragment.

NOTEBOOKS

Each student should keep a separate notebook. Do not count on your memory—write down your actions precisely in your notebook while you are doing the experiment. Your notebook should allow you to figure out why you did an experiment and exactly how you did it a long time after you did it. In addition, your notebook should be legible and thorough enough for someone else to read and understand what you did. The notebook should include:

a. *Rationale.* A short explanation of why you did the experiment.
b. *Protocol.* A detailed explanation of what you actually did. It is acceptable to cross-reference the lab manual, but you should record performance details that are not described in the lab manual (e.g., the genotype of mutants used and their auxotrophic requirements, any changes in the actual experiment compared to the protocol in the lab manual, and any unusual observations).
c. *Results.* The actual data (e.g., the number of colonies obtained, the OD determined, photos of agarose gels, etc.) should be included in your notebook. All plots and calculations should also be included. Show the equations used for all your calculations.
d. *Discussion.* A brief summary of the conclusions. (Did the controls work? What do the results mean?)

PROPOSALS

Since each group will be working with different mutants, sometimes each group will require different supplies. When this happens, I ask for a short description of the planned experiment (including controls) and the supplies needed. This allows us to obtain any special materials needed for the next experiments and to check for any weakness in the experimental design. Proposals should be turned in during the lab prior to when the supplies will be required.

PROGRESS REPORTS

I require each student to turn in a short report on each experiment. Progress reports should include a short summary and explanation of the experimental results. Progress reports should include the:

a. *Experiment number.*
b. *Purpose.* What did you expect to learn from the experiment?
c. *Results.* The results that should be included are outlined at the end of each experiment. When possible, the data should be presented in tables or figures. Indicate what controls were run and why.
d. *Conclusions.*

FINAL PAPER

I require a final paper summarizing the collective results of the experiments at the end of the course. It should be written up as a short paper suitable for submission to a scientific journal. You should compare and contrast your results with relevant references from the literature and cite the references in your paper.

LAB SAFETY

Safety is an important consideration for this type of course: the experiments involve bacteria, potentially harmful chemicals, and radioactivity. The instructor must thoroughly explain necessary safety precautions, demonstrate proper techniques, and carefully supervise the experiments.

BACTERIA

Many of the experiments in this course use *Salmonella typhimurium* LT2. Most natural isolates of *S. typhimurium* cause a serious bacteremia in mice and a less severe gastroenteritis in humans. Due to many years of maintenance in the lab, *S. typhimurium* LT2 is only weakly pathogenic for mice or humans, although very large doses may still cause an infection (see Sanderson and Hartman. 1978. *Bacteriol. Rev. 42*: 494). However, most of the strains used in this course contain plasmids or transposons that encode antibiotic resistance. In order to keep antibiotic resistant bacteria out of the environment (and you), all cultures and glassware that have come in contact with cultures should be sterilized after use and careful microbiological techniques should always be used when handling bacterial cultures.

RADIOACTIVITY

Several of the experiments in this course use radioactivity. Each step that involves radioactivity is indicated with an *. Proper procedures for the safe handling of radioactivity should be discussed in class and demonstrated. A few commonsense precautions should always be followed:

- Always wear gloves and a lab coat when handling radioactivity
- Never mouth-pipet a radioactive solution
- Only use radioisotopes in designated areas: no eating, drinking, or smoking is allowed in these areas
- Make sure all radioactive materials are labeled with a radiation sticker indicating the date, the isotope, and the activity
- If ^{32}P is substituted for ^{35}S in any of these experiments, always use ^{32}P behind a plexiglass shield
- Dispose of all contaminated materials and solutions in the appropriate radioactive waste containers
- Always check the work area for radioactivity when finished

CHEMICALS

Several chemicals used in this lab are hazardous. Precautions are noted in the lab manual when these chemicals are used. The following chemicals are particularly noteworthy:

- Phenol — can cause severe burns
- Acrylamide — potential neurotoxin
- Ethidium bromide — carcinogen

However, these chemicals are not harmful if used properly: always wear gloves when using potentially hazardous chemicals and never mouth-pipet them. If you accidentally splash any of these chemicals on your skin, *immediately rinse the area thoroughly with water* and inform the instructor. Discard the waste in appropriate biohazard containers. When in doubt, the toxicity of many chemicals can be looked up in the Merck Index.

ULTRAVIOLET LIGHT

Exposure to ultraviolet (UV) light can cause acute eye irritation. Since the retina cannot detect UV light, you can have serious eye damage and not realize it until 30 min to 24 hours after exposure. Therefore, *always wear appropriate eye protection* when using UV lamps.

ELECTRICITY

The voltages used for electrophoresis are sufficient to cause electrocution. Cover the buffer reservoirs during electrophoresis and place a "High Voltage" sign in front of the electrophoresis setup while it is running. Always turn off the power supply and unplug the leads before removing a gel.

LABELS

Since you will use common facilities, everything stored in an incubator, refrigerator, etc. must be labeled. In order to limit confusion, each group should select a unique set of initials which will serve as identification. You can use these initials together with a number to designate your mutant strains (e.g., AB1, AB2, etc.). This method will prevent duplication of strain names later. Always mark plates on the back (not the lids) with your initials, the date, and relevant experimental data (e.g., strain numbers).

GENETIC NOMENCLATURE

STRAIN COLLECTIONS

The ease of rapidly accumulating a large number of mutants requires careful bookkeeping to avoid confusing one mutant with another. Each mutant should be assigned a strain number. Strain numbers usually consist of two capital letters designating their source and a serial numbering of the strains in the collection of that laboratory.

GENOTYPES

An extensive genetic map is available for *S. typhimurium* (Sanderson and Roth, 1988) and *E. coli* (Bachmann and Low, 1980). The genes are named using standard genetic nomenclature. Each gene is assigned a three-letter designation, usually an abbreviation for the pathway or the phenotype of mutants. When the genotype is indicated, the three-letter designation is written in lower case. (For example, mutations affecting pyrimidine biosynthesis are designated *pyr*). Different genes that affect the same pathway are distinguished by a capital letter following the three-letter designation. (For example, the *pyrC* gene encodes the enzyme dihydroorotase and the *pyrD* gene encodes the enzyme dihydroorotate dehydrogenase).

Each mutation in the pathway is consecutively assigned a unique allele number. (For example, *pyrC19* refers to a particular *pyr* mutation that affects the *pyrC* gene. In order to distinguish each mutation, no other *pyr* mutation, regardless of the gene affected, will be assigned the allele number 19). A separate series of allele numbers is used for each three-letter locus designation. The entire genotype is italicized or underlined (e.g., *pyrC19*).

PHENOTYPES

It is often necessary to distinguish the phenotype of a strain from its genotype. The phenotype is usually indicated with the same three-letter designation as the genotype but phenotypes start with capital letters and are not underlined. (For example, strain TR251 [*hisC527 cysA1349 supD*] has a Cys⁺ His⁺ phenotype because the *supD* mutation suppresses the amber mutations in both the *cysA* and the *hisC* genes.)

TRANSPOSON INSERTIONS

Transposable elements can insert in known genes or in a site on the chromosome where no gene is yet known. When an insertion is in a known gene, the mutation is given a three-letter designation, gene designation, and allele number as described above, followed by a double colon then the type of insertion element. (For example, *pyrC691::Tn10* designates a particular insertion of the transposon Tn*10* within the *pyrC* gene).

When a transposon insertion is not in a known gene, it is named according to the map position of the insertion on the chromosome. Such insertions are named with a three-letter symbol starting with z. The second and third letters indicate the approximate map position in minutes: the second letter corresponds to 10-minute intervals of the genetic map numbered clockwise from minute 0 (a = 0-9; b = 10-19; c = 20-29, etc.); the third letter corresponds to minutes within any 10-minute segment (a = 0; b = 1; c = 2; etc). For example, a Tn*10* insertion located near *pyrC* at 23 minutes is designated *zcd* ::Tn*10* . Allele numbers are assigned sequentially to such insertions regardless of the letters appearing in the second and third positions, so if more refined mapping data suggests a new three-letter symbol, the allele number of the insertion mutation is retained. This nomenclature uses *zaa* (0 min) to *zjj* (99 min). Insertion mutations on extrachromosomal elements are designated with *zz*, followed by a letter denoting the element used. (For example, *zzf* is used for insertion mutations on an F ' plasmid.)

References

Bachmann, B., and K. Low. 1980. Linkage map of *Escherichia coli* K-12, Edition 6. *Microbiol. Rev. 44*: 1-56.

Davis, R., D. Botstein, and J. Roth. 1980. *Advanced Bacterial Genetics*, pp. 2-4. Cold Spring Harbor Laboratory, NY.

Sanderson, K., and J. Roth. 1988. Linkage map of *Salmonella typhimurium*, Edition VII. *Microbiol. Rev. 52*: 485-532.

BACTERIAL STRAINS USED

Strain	Genotype	Experiment
S. typhimurium		
LT2	Wild type	1,3,5
MS1063	*hisD9953* ::MudJ *his-9941* ::Mud1	1
MS1973	*del(his)640* /F′42(ts) *lac+ zzf-20*::Tn10 *finP301*	2
MS2	*thrA9 rpsL1*	2
MS4	*proA36 rpsL1*	2
MS5	*pyrC7 rpsL1*	2
MS6	*pyrF146 rpsL1*	2
MS8	*hisO1242 del(his)2236 rpsL1*	2
MS10	*ysA533 rpsL1*	2
MS12	*serA13 rpsL1*	2
MS13	*cysG439 rpsL1*	2
MS14	*metA53 rpsL1*	2
MS100	*ilv-508 rpsL1*	2
MS1974	*del(his)640* /F′42 *lac+ zzf-20*::Tn10 *finP301*	8
E. coli		
EM257	*del(recA-srl) srl* ::Tn10 *zfi* ::Tn10dCam *supF supE hsdR galK trp A metB lacY tonA* /pBR328 (*Hin* dIII fragment of MudJ)	5
EM158	*del(recA-srl) srl* ::Tn10 *zfi* ::Tn10dCam *supF supE hsdR galK trpR metB lacY tonA*	9
EM383	*hsd-5 del(lac-proBA) supE thi* /F′ *proB+ proA+ lacIq del(lac)M15*	10,11
EM258	*hsd-5 del(lac-proBA) supE thi* /F′ *proB+ proA+ lacIq del(lac)M15*/M13mp18	10
EM259	*hsd-5 del(lac-proBA) supE thi* /F′ *proB+ proA+ lacIq del(lac)M15* /M13mp19	10

[1]The *rpsL* mutation causes resistance to streptomycin (Str^r).

BASIC TECHNIQUES
FOR BACTERIAL GENETICS

STREAKING PLATES

When picking and streaking lots of bacterial colonies it is often quicker to use sterile toothpicks and sticks instead of using a wire loop that must be sterilized by passing it through a flame between each colony. Using the technique shown below, eight colonies can be streaked for isolation on a single plate. (Save the used toothpicks and sticks to be reautoclaved).

Divide the plate into 8 sectors.

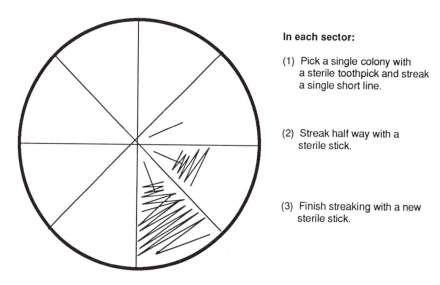

In each sector:

(1) Pick a single colony with a sterile toothpick and streak a single short line.

(2) Streak half way with a sterile stick.

(3) Finish streaking with a new sterile stick.

PICKING AND PATCH-ING COLONIES

When checking out bacterial strains, it is often useful to patch many colonies on a single petri plate so they can be tested simultaneously by replica printing onto a series of media. To do this, place a fresh master plate over a "patching grid." (A replica patching grid is supplied at the back of the manual). Touch the top of each colony with a sterile toothpick and draw a small x on the new master plate. (Only use each toothpick once. Save the used toothpicks to be reautoclaved). Many patches can be placed on a single plate. After patching incubate the plate overnight to let the patches grow. The next day this plate can be used as a master for replica printing. Always mark each plate at the top of the patch grid since the patch grids are symmetrical.

REPLICA PRINTING

This technique transfers cells from an array of colonies (or patches) on one plate, to a series of "replica" plates. Thus, each of the replica plates is inoculated by cells in the same arrangement as on the original ("master") plate. The transfer is done as follows:

1. Mark the top of the master plate and each replica plate. Always make the last plate replicated a control that all the colonies can grow on. This insures that failure of a colony to grow is not simply due to inefficient transfer.
2. Place a sterile velvet over the replica-printing block (fuzzy side up) and push a ring down over the velvet to hold it in place (see the figure below).

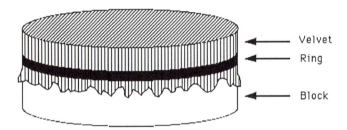

3. Press a plate containing an array of bacterial colonies (or patches) onto the surface of the sterile velvet. Press just hard enough that the fabric pattern become visible in the agar. Carefully lift the plate straight up and remove it to avoid smudging the print. Most of the cells on the plate will be transferred to the surface of the velvet.
4. Once a replica of the master plate has been formed on the velvet, press each of the fresh plates to be inoculated onto the surface of the velvet which carries cells. Many replicas can be made from a single master. (Save the velvets. If the velvets are washed and autoclaved, they can be used many times.)

PLATE MATINGS

1. Grow fresh overnight cultures of the donor (F⁺) and recipient (F⁻) strains. (If the F-factor is unstable it may be necessary to grow the donor in selective media. In addition, if the F-factor is temperature sensitive, then the donor must be grown at 30°C.)
2. Plate on a medium that selects for the phenotype of the exconjugants. Divide the plate into 3 sections.

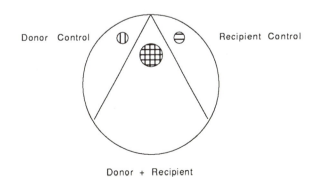

3. Place a small drop of the donor and recipient cultures on the plate as shown above. In the center section place the drops of the donor and recipient directly on top of each other. Allow the drops to dry onto the plate.
4. With a sterile stick, streak the plate from the drops out. (Save the used sticks to be reautoclaved.)
5. Incubate 1-2 days.

SPREADING PLATES

It is often necessary to spread a small amount of a supplement on the surface of a plate. This is done by using a spreader (a glass rod or Pasteur pipet bent to form an equilateral triangle which should be about 2/3 the diameter of the petri dish). Immediately prior to use, dip the spreader in a beaker of ethanol then pass it through a bunsen flame to ignite the alcohol. (Do not leave the spreader in the flame so long that it heats up). After the alcohol has "burned off," thoroughly spread the liquid on the surface of the plate until it is uniformly absorbed.

PHAGE P22

GROWTH OF P22

P22 is a temperate phage that infects *S. typhimurium* by binding to the O-antigen, part of the lipopolysaccharide on the outer membrane. After infection, P22 circularizes by recombination between terminal redundancies at each end of the phage DNA. During lytic growth, the circular genome of P22 initially undergoes several rounds of Oreplication, then changes to rolling circle replication. Rolling circle replication produces long concatemers of double stranded P22 DNA. These concatemers are packaged into phage heads by a "headful" mechanism: packaging is initiated at a specific sequence on the DNA called a Pac site, then a nuclease moves down the concatemer, cutting every 42 kb (Casjens and Hayden, 1988). Since the P22 genome is only 44 kb, this yields the terminal redundancy at the ends of P22 (Susskind and Botstein, 1978). This linear double stranded DNA is packaged into new phage particles. When the cell lyses, it releases 50-100 new phage.

P22 LYSATES

After P22 infects a cell and bursts it, the released phage infect other cells in the culture. Thus, after many rounds of phage multiplication and lysis of a bacterial culture, the broth contains a high concentration of phage. Unlysed cells and cell debris are removed by centrifugation, and chloroform ($CHCl_3$) is added to kill any remaining cells, yielding a solution of phage called a phage lysate. P22 lysates can be stored for many years at 4°C with some $CHCl_3$ at the bottom of the tube to keep it sterile. (If cells grow in the lysate, the phage will adsorb to them, drastically decreasing the titer of the phage stock.)

GENERALIZED TRANSDUCTION

There are sequences on the *S. typhimurium* chromosome that are homologous to the P22 Pac site. When P22 infects a cell, occasionally the P22 nuclease cuts one of these chromosomal sites and packages 44 kb fragments of chromosomal DNA into P22 phage heads. The P22 particles carrying chromosomal DNA (transducing particles) can inject this DNA into a new host. The DNA can then recombine into the chromosome by homologous recombination. Since P22 can transfer DNA fragments from all regions of the chromosome, this process is called generalized transduction (Masters, 1985; Margolin, 1987).

In this course we use P22 HT105/1 *int-201* , a P22 mutant that is very useful for generalized transduction. This phage has a high transducing (HT) frequency due to a nuclease with less specificity for the Pac sequence. About 50% of the P22 HT phage heads carry random transducing fragments of chromosomal DNA (Schmeiger, 1972). The *int* mutation prevents formation of stable lysogens.

Unlike many phage, P22 is very "user friendly." P22 can infect overnight cultures, so you do not have to monitor carefully the growth of the culture to catch it in log phase. P22 nearly always produces high titer phage stocks (10^{10}-10^{11} phage/ml).

Finally, P22 HT produces such a high percentage of transducing particles that it is easy to look for rare events, like recombination between very close genetic markers (Sanderson and Roth, 1983).

References

Casjens, S. and M. Hayden. 1988. Analysis *in vivo* of the bacteriophage P22 headful nuclease. *J. Mol. Biol. 199*: 467-474.

Margolin, P. 1987. Generalized transduction. In Neidhardt, F., J. Ingraham, K. Low, B. Magasanik, M. Schaechter, and H. Umbarger. Escherichia coli *and* Salmonella typhimurium: *Cellular and Molecular Biology.* American Society for Microbiology, Washington, D.C.

Masters, M. 1985. Generalized transduction. In J. Scaife, D. Leach, and A. Galizzi (eds), *Genetics of Bacteria*, pp. 197-205. Academic Press, NY.

Poteete, A. 1988. Bacteriophage P22. *In* R. Calender (ed.), *The Bacteriophages*, pp. 647-682. Plenum, NY.

Sanderson, K., and J. Roth. 1983. Linkage Map of *Salmonella typhimurium*, Edition VI. *Microbiol. Rev. 47:* 410-453.

Schmieger, H. 1972. Phage P22 mutants with increased or decreased transduction activities. *Mol. Gen. Genet. 119*: 75-88.

Susskind, M., and D. Botstein. 1978. Molecular genetics of bacteriophage P22. *Microbiol. Rev. 42:* 385-413.

**A. PREPARATION OF
P22 PHAGE LYSATES**

1. Grow an overnight culture of the donor strain in NB.
2. Add 4 ml P22 broth to 1 ml of the overnight culture. The final multiplicity of infection (moi) should be about 0.01-0.1 pfu/cell.
3. Incubate 8-16 hrs in a 37°C shaker (temperature sensitive mutants can be grown at 30°C).
4. Spin down 20 min in a clinical centrifuge to pellet the cell debris.
5. Pour the supernatant into a screw capped test tube. Add several drops of chloroform and vortex. Store at 4°C (a good lysate should contain 10^{10}-10^{11} pfu/ml).

P22 Broth:

Mix the following solutions:
    100 ml sterile NB
      2 ml 50x E medium (sterilized with $CHCl_3$)
      1 ml sterile 20% glucose
    0.1 ml P22 HT int phage (sterilized with $CHCl_3$)
Store at 4°C.

**B. P22
    TRANSDUCTIONS**

1. Grow the recipient strain overnight in 2 ml NB.
2. Dilute the phage 1/50 in sterile 0.85% NaCl just before use. (If the phage titer is unusually low or high you may need to use a different dilution.)
3. Mix the cells and diluted phage directly on selection plates as follows:

Plate	ml cells	ml phage	
A	0.1	—	cell control
B	0.1	0.05	
C	0.1	0.10	
D	0.1	0.20	
E	—	0.20	phage control

4. Spread the plates with an alcohol-flamed glass spreader.
5. Incubate the plates upside-down.
6. Count the colonies on each plate. There should be no growth on the cell or phage control plates. Any colonies you plan to save should be purified on green plates (see pages 23-25).

NOTE: For certain transductions (for example, when selecting Kan^r or Str^r) phenotypic expression is required before plating on the selective medium. Phenotypic expression can be done in two ways:

• Mix cells and phage in a test tube and incubate about 1 hr before plating on the selective medium.

• Spread the cells and phage on nonselective medium (e.g., NB plates), incubate 4-8 hrs, then replicate onto the selective plates. This method usually gives more colonies and ensures that different colonies are not due to siblings.

C. SPOT TITERING P22 LYSATES

When P22 HT int is used, phage stocks do not usually need to be titered. However, it is a good idea to check the titer of your phage stock if a transduction does not work.

1. Divide a green plate into 4 sectors with a marking pen. (Make sure the surface of the plate is not noticeably wet. If the plate is wet, place it in an incubator or oven with the lid ajar until dry.)
2. Melt TS top agar in a microwave. For each phage to be titered, add 2.5 ml of the melted top agar to a test tube and place in a 50°C heating block. After the top agar cools to about 50°C, add 0.1 ml of an overnight culture of *S. typhimurium* LT2 to each tube.
3. Immediately swirl and pour onto the green plate.
4. Allow the top agar to solidify for 15-30 min.
5. Spot 20 µl of appropriate phage dilutions onto the lawn (usually 10^{-6}, 10^{-7}, 10^{-8}, 10^{-9} dilutions in sterile 0.85% NaCl).

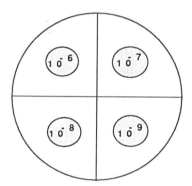

6. Leave the plate on the bench for about 30 min to allow the drops of phage to dry.
7. Incubate the plates upside-down at 37°C overnight.
8. Count the number of plaques in each spot. Each viable phage (plaque forming unit or pfu) will produce one plaque. Calculate the phage titer as follows:

$$\text{pfu / ml} = \frac{\text{number plaques x dilution factor}}{20\ \mu l} \times \frac{1000\ \mu l}{ml}$$

D. CHECKING *S typhimurium* for P22 SENSITIVITY

In addition to receiving a transducing fragment, some of the transductants may have also been infected with P22 phage. When *S. typhimurium* is infected with phage P22 the following pathways are possible (Bochner, 1984):

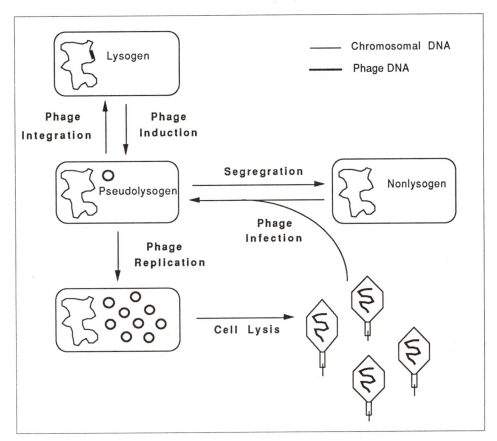

Pseudolysogens can be differentiated from nonlysogens and true lysogens on green plates. Green plates contain pH indicators which are green at neutral pH but turn dark blue at low pH. When streaked on green plates, nonlysogens and true lysogens form light-colored colonies. However, in a colony containing pseudolysogens many cells are undergoing lysis which lowers the pH of the medium resulting in dark blue colonies. Since in pseudolysogens the phage exists as a low copy number plasmid, it is possible to obtain "phage-free" segregants by streaking for isolated colonies on green plates. The "phage-free" segregants will form light-colored colonies while the pseudolysogens will remain blue.

P22 HT int phage. P22 HT *int* phage is used for transductions. The *int* mutation prevents formation of stable lysogens. However, when cells are left on plates with lytic phage, there is a strong selection for revertants that form stable lysogens. Since stable lysogens cannot be reinfected with P22, such transductants are not very useful for genetic studies. Therefore, it is important to "clean up" transductants on green plates as soon as possible after a transduction. Ca++ is required for phage adsorption. EGTA chelates Ca++ so it is not available for phage adsorption. Therefore, we sometimes include EGTA in plates to decrease reinfection of transductants. (EGTA cannot be added to the initial selection plates or it will prevent adsorption of the transducing phage also). However, even colonies from EGTA plates must be cleaned up on green plates. Once light-colored colonies have been purified from green plates, they should be checked to make sure they are not true lysogens. This is done by cross-streaking against a P22 *c2* mutant called H5. (The P22 *c2* gene encodes a repressor equivalent to *cI* of phage lambda. Thus, P22 *c2* mutants produce clear plaques). Phage free cells are infected by H5 and lysed (H5 sensitive) but P22 lysogens are not infected by H5 (H5 resistant). The test for H5 sensitivity is done as shown on the next page.

NOTE. Be careful not to "dig into" the agar when streaking green plates: cells ferment more glucose when anaerobic so all of the colonies will appear dark blue. Also, do not store colonies on green plates: when left on green plates for many days, all of the colonies will turn dark colored and some mutants may be killed by the accumulated fermentation products.

Checking strains for sensitivity to phage P22

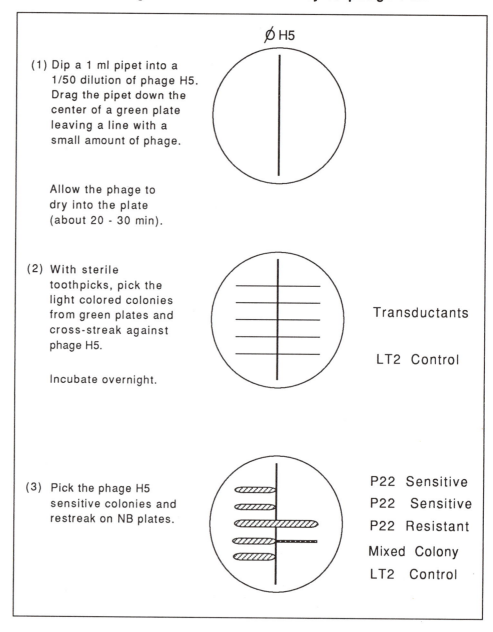

(1) Dip a 1 ml pipet into a 1/50 dilution of phage H5. Drag the pipet down the center of a green plate leaving a line with a small amount of phage.

Allow the phage to dry into the plate (about 20 - 30 min).

ⵁ H5

(2) With sterile toothpicks, pick the light colored colonies from green plates and cross-streak against phage H5.

Incubate overnight.

Transductants

LT2 Control

(3) Pick the phage H5 sensitive colonies and restreak on NB plates.

P22 Sensitive

P22 Sensitive

P22 Resistant

Mixed Colony

LT2 Control

References

Bochner, B. 1984. Curing bacterial cells of lysogenic viruses by using UCB indicator plates. *Biotechniques* , Sept / Oct pp. 234-240.

Davis, R., D. Botstein, and J. Roth. 1980. *Advanced Bacterial Genetics*, pp. 16-18. Cold Spring Harbor Laboratory, NY.

1

CONSTRUCTION OF OPERON FUSIONS

The purpose of this experiment is to isolate auxotrophic transposon insertion mutants of *S. typhimurium*. Mu is a transposon that can insert at essentially random sites in the *S. typhimurium* chromosome. When Mu inserts into a gene it disrupts the gene and hence causes a mutation. Mu insertions are also polar on downstream genes in an operon. If Mu inserts into a gene required for biosynthesis of an essential metabolite, the mutant will be an auxotroph: it will require supplementation with the missing metabolite.

MUD FUSIONS

Derivatives of Mu have been constructed that carry the *lac* operon (without its promoter) near one end of Mu (Casadaban and Cohen, 1979; Castilho et al., 1984). When these Mu derivatives (Mud) are inserted in a gene in the correct orientation, the *lac* genes are expressed from the promoter of the mutated gene. Hence, expression of the *lac* operon is directly proportional to expression of the mutated gene. Since expression of the *lacZ* gene can be easily detected on indicator plates and quantitated by assaying ß-galactosidase activity, the expression of the mutant gene can be easily studied *in vivo*. Two types of Mud fusion vectors are available.

(1) In the first type of Mud vector the lac genes carry their own translational start sites, so expression of *lacZ* is directly proportional to transcription from the promoter of the mutated gene but translation of *lacZ* is independent of the mutated gene. These vectors form operon fusions.

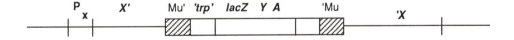

(2) In the second type of Mud vector the lac genes have no translational start sites, so expression of *lacZ* is determined by both the transcription and translation of the mutant gene. These vectors form gene fusions. The gene fusion vectors produce a hybrid protein with the N-terminus of the mutated gene and the C-terminus of ß-galactosidase.

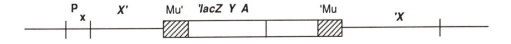

MUD NOMENCLATURE

A variety of phage Mu derivatives (Mud phages) have been used as transposons for formation of operon and gene fusions. Several useful Mud vectors are shown in Figure 1. It is possible to substitute one type of Mu derivative for another by recombination events within the Mud phage that do not alter the insertion site or the flanking sequences in the host chromosome. This is a useful genetic trick but it causes some nomenclature problems. Since a variety of Mud phages are available and the names fit no apparent uniform system, Hughes and Roth proposed a simplified nomenclature for Mud insertions. If simple mutations are added to these phages, the genotype can be added to the insertion designation. For example, a Tn5 insertion in the Amp^{r}gene can be recombined onto any of the Mud phages carrying Amp^r: if *his-9897*::Mud1 is transduced with the Tn5 in Amp^r it would be called *his-9897*::Mud1(*bla*::Tn5). The allele number of a Mud insertion is not changed when the Mud is altered by recombination with other Mud phages. For example, *his–9897*::Mud1 retains this allele number when it is replaced with a Kan^r Mud. It becomes *his-9897*::MudJ.

MUD INSERTIONS

In this experiment we will isolate mini-Mud operon fusions using a Kan^r mini-Mud derivative called MudJ. MudJ can insert into a gene in two orientations as shown below. (A) If it is inserted in the correct orientation (i.e., the *lac* operon faces the same direction the gene is transcribed), a MudJ operon fusion is formed and the *lac* operon is expressed. (B) If it is inserted in the opposite orientation, transcription from the gene runs into the wrong end of the Mud so the *lac* operon is not expressed.

MudJ operon fusions can be identified on plates containing the indicator Xgal. Xgal (5-bromo-4-chloro-3-indoyl-ß-D-galactoside) is a colorless lactose analog that is a very sensitive indicator of ß-galactosidase activity. When Xgal is cleaved by ß-galactosidase, an insoluble deep blue dye is produced.

TRANSPOSITION

MudJ does not carry the Mu transposition functions. However, transposase provided in *trans* from Mud1 can promote transposition of the MudJ. Transposition of MudJ can be selected using a genetic trick called "transitory cis-complementation" (Hughes and Roth, 1988). Phage P22 can carry about 44 kb of DNA. Mud1 is about 37 kb and the mini-Mud derivatives are about 10 kb. Hence, P22 cannot package a complete Mud1 and mini-Mud in the same phage head. The donor strain has a backwards Mud1 insertion in the *hisA* gene (i.e., the Mud1 insertion is Xgal$^-$) and a Xgal$^+$ insertion of MudJ in the *hisD* gene (see the diagram opposite).

Figure 1. Some Mud fusion vectors

ORIGINAL NAME	ABBREVIATED NAME	SIZE (kb)	
Mud-I	Mud1	37.2	Original operon fusion phage (Casadaban and Cohen, 1979)
Mud11301	Mud2	35.6	Similar to Mud1 but forms gene fusions (Casadaban and Chou, 1984)
Mud1-8	MudA	37.2	Derivative of Mud1 with suppressible transposition defects (Hughes and Roth, 1984)
Mud2-8	MudB	35.6	Derivative of Mud2 with suppressible transposition defects (Hughes and Roth, 1984)
Mud11734	MudJ	11.3	Operon fusion min-Mud deleted for transposition functions (Beatriz et al., 1984)
Mud11734	MudK	9.7	Gene fusion mini-Mud deleted for transposition functions (Beatriz et al.,1984)

In this strain the end of Mud1 with the Mu genes required for transposition (labeled Tpn) are closest to the MudJ insertion (Symonds et al., 1987). Phage P22 grown on this strain can be used to transduce MudJ into another strain by selecting Kan^r. There are two ways MudJ can be inherited.

(1) Some P22 particles will package the MudJ and the transposase genes from the Mud1. When such transducing fragments enter a new cell, the transposase genes are expressed and can promote transposition of the MudJ. Transposition of MudJ into a new site in the chromosome yields His$^+$ Kan^r colonies. The Mud1 fragment cannot be coinherited since both ends of the Mud1 are required for transposition, but they will not fit into the same phage head with the MudJ. Hence the Mud1 remains on an "abortive transducing fragment" that is not stably incorporated into the chromosome. Any rare colonies that coinherited the Mud1 will be Amp^r. The copy of MudJ that "hopped" into the chromosome is stable since it lacks the Mu transposase genes. Usually about 50% of the MudJ transductants selected on rich medium have transposed into new genes.

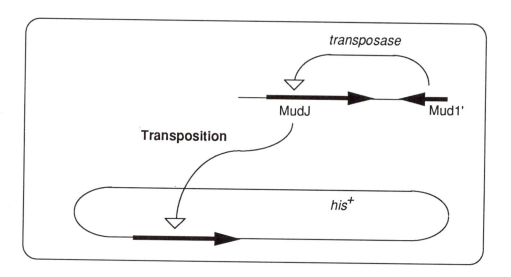

(2) Alternatively, the MudJ can be inherited by homologous recombination into the *hisD* gene. Homologous recombination yields His$^-$ Kan^r colonies. Since homologous recombination just regenerates the original MudJ insertion, we are not interested in isolating the His$^-$ Kan^r colonies. Such mutants can be avoided by selecting for Kan^r colonies on plates that contain a mixture of many potential auxotrophic supplements except histidine — *hisD*::MudJ insertions require histidine so they will not grow without histidine. However, even on medium without added histidine some His$^-$ mutants may form small colonies due to cross-feeding of histidine from nearby His$^+$ colonies, so it is important to double-check the phenotype of all potential auxotrophic mutants.

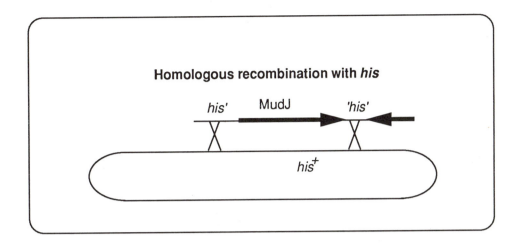

References

Beatriz, A. C., Olfson, P., and Casadaban, M. J. 1984. Plasmid insertion mutagenesis and *lac* gene fusion with mini-Mu bacteriophage transposons. *J. Bacteriol.* 158: 488-495.

Casadaban, M. J., and J. Chou. 1984. *In vivo* formation of gene fusions encoding hybrid ß-galactosidase proteins in one step with a transposable Mu-*lac* transducing phage. *Proc. Natl. Acad. Sci. USA.81*: 535-539.

Casadaban, M. J., and S. N. Cohen. 1979. Lactose genes fused to exogenous promoters in one step using a Mu-*lac* bacteriophage: *In vivo* probe for transcriptional control sequences. *Proc. Natl. Acad. Sci. USA 76*: 4530-4533.

Castilho, B., P. Olfson, and M. Casadaban. 1984. Plasmid insertion mutagenesis and *lac* gene fusion with mini-Mu bacteriophage transposons. *J. Bacteriol. 158*: 488-495.

Hughes, K., and J. Roth. 1984. Conditionally transposition-defective derivative of MudI (Amp, Lac). *J. Bacteriol. 159*: 130-137.

Hughes, K., and J. Roth. 1988. Transitory cis-complementation: a general method for providing transposase to defective transposons. *Genetics 119*: 9-12.

Symonds, N., A. Toussaint, P. van de Putte, and M. Howe. 1987. *Phage Mu.* Cold Spring Harbor Laboratory, NY.

Wilson, R., and S. Maloy. 1987. Isolation and characterization of *Salmonella typhimurium* glyoxylate shunt mutants. *J. Bacteriol. 169*: 3029-3034.

1A. Isolation of MudJ Insertion Mutants

1. Check out the phenotypes of the strains:
   LT2 (wild type) should be His$^+$, Kan^s, and Amp^s;
   MS1063 (*hisD9953*::MudJ *his-9941*::Mud1) should be His$^-$, Kan^r, and Amp^r.
   Always grow MS1063 at 30°C because Mud1 has a temperature sensitive repressor.

2. Grow a P22 HT *int* phage lysate on MS1063 at 30°C (see page 13). Start an overnight culture of LT2 in NB.

3. Dilute the P22 lysate grown on MS1063 1/50 in T2 buffer. Transduce LT2 with the diluted lysate as shown below. Selection for Kan^r requires phenotypic expression: spread cells and phage on NB plates.

Plate	ml LT2	ml P22	
A	0.1	—	Cell control
B	—	0.2	Phage control
C	0.1	0.2	
D	0.1	0.2	
E	0.1	0.2	
F	0.1	0.2	
G	0.1	0.2	

4. Incubate the plates upside-down for 3-4 hr at 37°C, then replicate onto glucose + 125 µg/ml Kan + EGTA + Xgal plates containing all of the pool supplements except histidine. Incubate overnight at 37°C.

5. Check the plates. There should be no growth on either of the control plates. Replica plate the 5 transduction plates onto the following media to screen for insertion mutants.

   NB + Amp
   E + Glucose + Kan
   E + Glucose + Kan + His
   E + Glucose + Kan + EGTA + Xgal + Pool supplements

6. Incubate the master plate and the replica plates overnight at 37°C.

7. Note the number of colonies that grow on each plate. Also note the number of colonies that are blue on the plate containing Xgal.

8. Auxotrophs caused by transposition of MudJ will grow on the master plates but not on E + His + Kan plates or the E + Kan plates. Pick these His$^+$ auxotrophic transductants and streak for single colonies on Green plates (8 colonies per plate). Incubate overnight at 37°C.

9. From each streak, pick a well isolated light colored colony and restreak another green plate. Incubate overnight at 37°C.

10. From each streak, pick a well isolated light-colored colony and cross-streak against phage H5 on a green plate to check for phage sensitivity (see page 16). Incubate overnight at 37°C.

Preparation of Supplement Pools and Plates

Mixed-pool supplements:

Mix together 10 ml of each of the sterile pool supplements except histidine (see Appendix 4). The solution of mixed supplements can be filter sterilized if necessary.

E + Glucose + Kan + Xgal + EGTA + Pool plates:

Add 15 g agar to 500 ml dH$_2$O. In a separate flask add 20 ml 50x E medium to 170 ml dH$_2$O. Autoclave both flasks.

After autoclaving, combine the agar with the E medium, allow them to cool to about 50°C, then add:

125	mg	Kanamycin SO$_4$
10	ml	sterile 20% glucose
10	ml	sterile 1 M EGTA
1	ml	20 mg/ml Xgal dissolved in dimethylformamide
310	ml	sterile mixed-pool supplements

Results - Experiment 1A

Plate #	# Kan^r Colonies	# Amp^r Colonies	# Auxotrophic Colonies	Comments
1				
2				
3				
4				
5				
6				
7				

1B. Characterization of Auxotrophic Requirements

Auxotrophic mutants can be easily screened for their nutritional requirements by testing for growth on crossed-pool plates (Davis, Botstein, and Roth, 1980). Eleven pools contain the common auxotrophic supplements that account for most of the major biosynthetic pathways. Each amino acid or nucleotide is present in two pools of the eleven. A mutant requiring one amino acid or nucleotide would grow only on the two pools which contain it. For example, methionine is in pool 4 and pool 6 so a *met* auxotroph would only grow on plates 4 and 6. Some auxotrophs require two supplements: these mutants will only grow on the one pool that contains both supplements. For example, an *ilv* mutant requires isoleucine and valine so it would grow only on pool 7. Some other potential auxotrophic requirements are only present in pool 11. It should be possible to identify most of your auxotrophs by this method. However, some mutations may cause unusual pleiotropic mutations that may not grow on any of the pool plates.

1.   Pick phage sensitive colonies of potential auxotrophs and patch onto E + glucose + Kan + EGTA + Xgal + pool supplements (minus histidine) plates. Also patch LT2 and MS1063 controls. Incubate overnight at 37°C.
2.   Examine the plates. All of the patches should grow except the controls: LT2 should not grow because it is Kan^s and MS1063 should not grow because it is His$^-$.
3.   Replica plate onto E + glucose plates with each of the "crossed pool" supplements (see the table on the next page). Incubate plates at 37°C.
4.   Score the plates and determine the auxotrophic requirements using the table on the next page.
5.   Double-check the auxotrophic requirements by streaking each mutant on an E + glucose plate and an E + glucose plate spread with 0.1 ml of the requirement. Incubate the plates overnight at 37°C.
6.   Assign a strain number to each auxotrophic mutant that has the expected phenotype ("checks out") and streak it on an NB plate. Store the plate in the refrigerator.
7.   In order to avoid inadvertantly losing your mutants, you can also store a frozen culture. Grow an overnight culture of each auxotroph in 2 ml NB. Add about 1 ml to a sterile vial containing 0.2 ml DMSO, label the vial, then place it in the freezer at -70°C.

The composition of the auxotrophic pool plates is shown in the following table (Davis, Botstein, and Roth, 1980). The supplements in pools 1-5 are listed in the vertical columns and the supplements in pools 6-10 are listed in horizontal rows. Pool 11 contains supplements (mainly vitamins) not included in the other pools; its contents are listed in the horizontal row at the bottom of the table. Some comments on interpreting the results are listed below the table on the following page.

Crossed pool plates:

	1	2	3	4	5		
6	ade	gua	cys	met	thi		
7	his	leu	ile	lys	val		
8	phe	tyr	trp	thr	pro	PABA	DHBA
9	gln	asn	ura	asp	arg		
10	thy	ser	glu	DAP	gly		
11	Pyridoxine, nicotinic acid, biotin, pantothenate, ala						

a. Some purine mutants grow on adenosine or guanosine (*purC*, *purE*, or *purH*); these mutants will grow on pools 1, 2, and 6.

b. Some purine mutants require adenosine + thiamine (*purD*, *purF*, *purG*, or *purI*); these mutants will only grow on pool 6.

c. *pyrA* mutants require uracil + arginine; they only grow on pool 9.

d. Mutants requiring isoleucine + valine (*ilv*) only grow on pool 7.

e. Mutants with early blocks in the aromatic pathway will only grow on pool 8. In addition to the aromatic amino acids, pool 8 contains the precursors p-aminobenzoic acid (PABA) and dihydroxybenzoic acid (DHBA).

f. Mutants with early blocks in the lysine pathway require lysine + diaminopimelic acid (DAP); they will only grow on pool 4.

g. Some *thi* mutants require very small amounts of thiamine; they often pick up enough thiamine from pools 5 and 6 to grow on all subsequent pool plates.

h. Some mutants require either cysteine or methionine (see *J. Gen. Microbiol.* 1975. *89*:353); these mutants will grow on pools 3, 4, or 6.

i. Note that there is insufficient glutamine in NB to supplement mutants that require high glutamine (e.g. *glnA* mutants).

j. TCA cycle mutants may have complex requirements.

k. Solutions of the nutrient pools (1-11) can be made up by mixing equal volumes of the nutrients contained in each pool (see Appendix 4). Use 25 ml of the "pool solution" per liter of medium.

Many of the biosynthetic pathways are reviewed in Neidhardt et al. (1987) and Herrmann and Somerville (1983). The *S. typhimurium* genetic map and a list of the known genes can be found in Sanderson and Roth (1988).

References

Davis, R., D. Botstein, and J. Roth. 1980. *Advanced Bacterial Genetics*, pp. 15-18. Cold Spring Harbor Laboratory, NY.

Herrmann, K. and R. Somerville. 1983. *Amino Acids: Biosynthesis and Genetic Regulation*. Addison-Wesley, Reading, MA.

Neidhardt, F., J. Ingraham, K. Low, B. Magasanik, M. Schaechter, and H. Umbarger. 1987. Escherichia coli *and* Salmonella typhimurium: *Cellular and Molecular Biology*. American Society for Microbiology, Washington, D.C.

Sanderson, K., and J. Roth. 1988. Linkage map of *Salmonella typhimurium*, edition VI. *Microbiol. Rev.* 52: 485-532.

Results - Experiment 1B

Strain	Growth on Pool											Interpretation	Confirmation		Comments
	1	2	3	4	5	6	7	8	9	10	11		Supplement	Growth	

MAPPING OPERON FUSIONS

The previous experiment involved isolation of mini-Mud insertion mutants and identification of the mutant phenotype. Since many pathways have genes that map at several positions on the chromosome, identification of the mutated gene requires mapping the mutation. The purpose of this experiment is to determine where the MudJ insertion mutation maps on the *S. typhimurium* chromosome. The mutations can be mapped genetically by matings (Experiment 2A) and transductions (Experiment 2B).

Hfr MAPPING

By using the MudJ insertion as a "portable region of homology," an Hfr can be constructed with its origin within the insertion (Chumley et al., 1979; Maloy and Roth, 1983). The origin of the Hfr can be mapped by mating with known auxotrophic mutants, thus identifying the map location of the original MudJ insertion.

ISOLATION OF AN Hfr

An F' derivative of the MudJ mutant must be isolated before an Hfr can be made (Figure 2-1). First, an F' ts Tn*10 lac* + is mated into the MudJ recipient. The MudJ/ F'tsTn*10 lac* + exconjugant is obtained by selecting for growth on minimal Tet plates at 30°C. The donor is His⁻ so it cannot grow on minimal medium without histidine, and the recipient is Tetˢ so it cannot grow on medium containing tetracycline. Thus, only the MudJ / F'ts Tn*10 lac* + exconjugants will grow on these plates. Since replication of the F' is temperature sensitive, the matings must be done at 30°C.

MAPPING THE Hfr

Once the MudJ / F'ts Tn*10 lac* + exconjugants have been constructed, an Hfr can be selected as shown in Figure 2-2. Homologous recombination between the *lac* genes on the F' and the *lac* genes in the MudJ operon fusion will integrate the F' into the chromosome, forming an Hfr. It is possible to select for such Hfr's by growing the MudJ / F'ts Tn*10 lac* + exconjugants on minimal Tet medium at 43°C. Replication of the F' is temperature sensitive, so it cannot replicate at 43°C. Therefore, most of the cells will lose the F' becoming Tetˢ. However, if the F' integrates into the chromosome to form an Hfr, the cells will remain Tetʳ.

Other than the MudJ mutation, the Hfr formed is wild type for all the other genes. When mated with auxotrophic mutants, the frequency of transfer of the wild type allele is usually proportional to the distance between the origin of transfer and the gene selected: if a gene is close, it will be transferred at high frequency and if it is farther away it will be transferred at a lower frequency (Miller, 1972). (However, sometimes genes that are very close to the origin of transfer are are not inherited at the predicted frequency, probably due to the nonhomology between the donor and recipient in this region.) By mating the Hfr with a known set of auxotrophic mutants, the approximate map location of the original MudJ insertion can be determined (Figure 2-3). With the recipients used in this experiment, most mutations can be mapped within a 10 min region of the chromosome (Sanderson and Roth, 1988).

DIRECTION OF TRANSCRIPTION

Since the direction of transfer of the Hfr is determined by the *lac* homology and since *lac* is transcribed from the promoter of the mutant gene, this technique also indicates the direction of transcription of the operon fusion with respect to the *S. typhimurium* chromosome (Maloy and Roth, 1983).

References

Chumley, F., R. Menzel, and J. Roth. 1979. Hfr formation directed by Tn*10*. *Genetics 91*: 639-655.

Maloy, S., and J. Roth. 1983. Regulation of proline utilization of *Salmonella typhimurium* : characterization of *put* ::Mud(Ap,*lac*) operon fusions. *J. Bacteriol. 154:* 561-568.

Miller, J. 1972. *Experiments in Molecular Genetics*, pp. 63-81. Cold Spring Harbor Laboratory, NY.

Sanderson, K., and J. Roth. 1988. Linkage map of *Salmonella typhimurium*, Edition VI. *Microbiol. Rev. 52*: 485-532.

Figure 2-1. Isolation of F' derivatives of MudJ insertion mutants

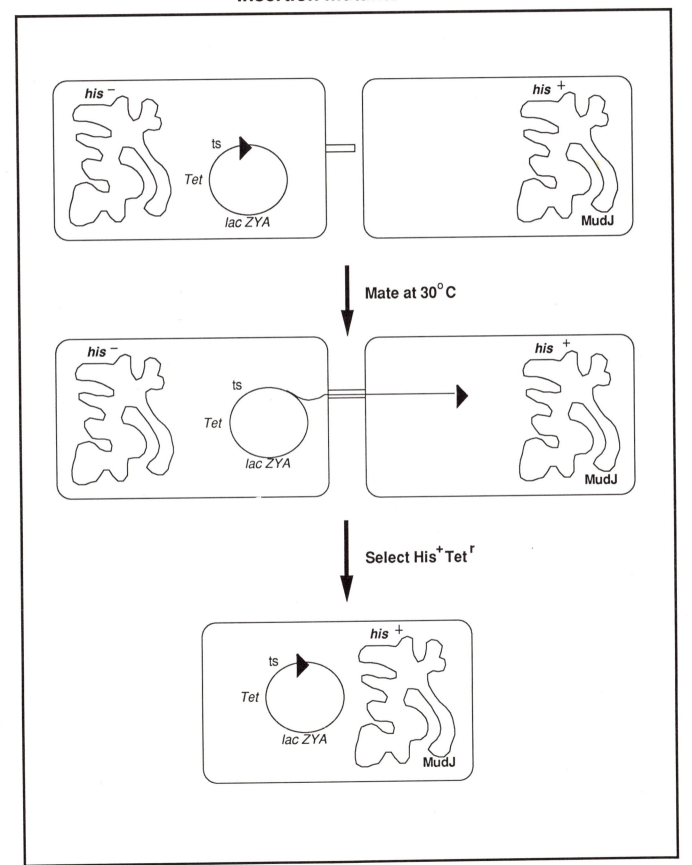

Figure 2-2. Selection for an Hfr by homologous recombination between *lac* genes on the F' and the MudJ insertion

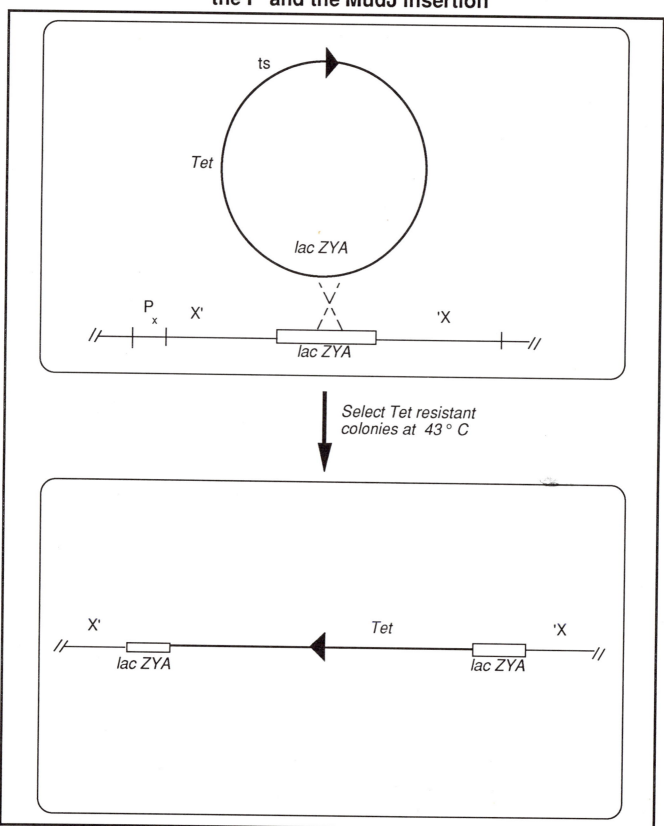

Figure 2-3. Hfr mapping

The probability that a gene will be transferred from an Hfr to a recipient depends upon its position and distance relative to the Hfr. For example, in the diagram below the genes Z,Y, and X would be transferred frequently and the genes C, B, and A would only be transferred very rarely.

Therefore, the origin of a Hfr can be mapped on the *S. typhimurium* chromosome by mating with known auxotrophic recipients. If the auxotrophic mutation maps close to the origin of the Hfr, many wild type recombinants will be recovered. The farther the mutation is from the origin of the Hfr, the fewer the number of wild type recombinants. Thus, by using mutants in the genes shown below, the origin of an Hfr can be mapped within about 10 min of the genetic map.

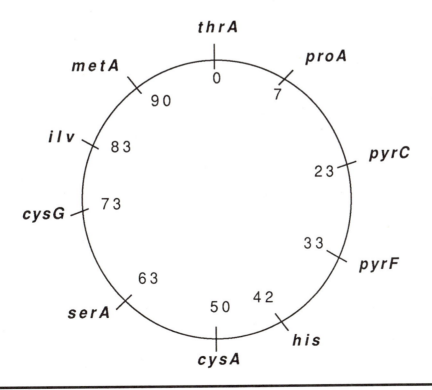

2A. Hfr Mapping

Isolation of the MudJ Hfr donor:

1. Grow the donor strain MS1973 [*del(his)* /F'42(ts) *lac* $^+$ *zzf-20* ::Tn*10*] on E + glucose + Tet + His plates at 30°C.
2. Subculture donor strain and the MudJ recipient strains into 2 ml NB. Grow overnight in 30°C shaker.
3. Add the donors and recipients to sterile microfuge tubes as follows:

Tube	Donor	Recipient	
A	0.5 ml MS1973	—	MS1973 control
B	0.5 ml MS1973	0.5 ml MudJ	
C	—	0.5 ml MudJ	MudJ control

4. Incubate the tubes at least 2 hr in the 30°C incubator.
5. Centrifuge for 20 sec in the microfuge to pellet the cells.
6. Pour off the supernatant and resuspend the cell pellet in 0.1 ml sterile 0.85% NaCl.
7. Spread two E + glucose + Tet plates with 0.1 ml of the appropriate auxotrophic supplement.
8. Divide the plate into 3 sections. Place a small drop of the resuspended cells on the plate as shown on page 17. Allow the drops to dry onto the plate (usually takes about 20 min).
9. With a sterile stick, streak the plate from the drops out. Incubate the plates upside-down for 2 days at 30°C.
10. There should be no growth in either of the controls and single colonies in the mating spot. Pick isolated large colonies and restreak onto E + glucose + Tet plates spread with 0.1 ml of the appropriate auxotrophic supplement. Incubate 1-2 days at 43°C.
11. Subculture three isolated large colonies from the 43°C plate into separate tubes containing 2.5 ml E + glucose + Tet liquid medium and the appropriate auxotrophic supplement. (Small colonies are often not true exconjugants and will fail to grow when transferred to liquid medium.)
12. Grow 2-3 days on a 43°C shaker. These cultures will contain the MudJ Hfr donors.

Determining the map position of the Hfr insertion:

1. Grow up the following Str^r auxotrophic recipient strains in 2.5 ml NB at 37°C.

   S. typhimurium LT2 Str^r auxotrophic recipients:

STRAIN	AUXOTROPHIC MARKER	MIN
MS2	*thrA*	0
MS4	*proA*	7
MS5	*pyrC*	23
MS6	*pyrF*	33
MS8	*his*	42
MS10	*cysA*	50
MS12	*serA*	63
MS13	*cysG*	73
MS100	*ilv*	83
MS14	*metA*	90

2. Mate the 43°C liquid cultures (MudJ Hfr donors) with the auxotrophic recipients:
   a. Mix each Str^r auxotrophic recipient with the MudJ Hfr donor in separate sterile microfuge tubes as shown below.

Tube	Donor cells	Recipient cells	
A	0.2 ml	—	Donor control
B	—	0.2 ml	Recipient control
C	0.2 ml	0.2 ml	

   b. Incubate at least 2 hr at 37°C.
   c. Centrifuge for 20 sec in a microfuge to pellet the cells.
   d. Pour off the supernatant. Resuspend the cells in 0.2 ml sterile 0.85% NaCl.
   e. Spread 0.1 ml of each mating mixture on separate E + 0.2% glucose + 1% NB + 2 mg/ml Str plates. Str^r selects against the donor and growth on minimal medium selects against the auxotrophic recipients. (Requires 12 plates: 10 recipients + 1 control with only donor cells +1 control with only recipient cells).

3.   Count the number of colonies on each plate. From the gradient of transfer and the known map position of the auxotrophic recipients determine the map position of your Mud insertion.

Using this procedure most mutations can be mapped within a 10 min region of the chromosome. Using known mutations in this region of the *S. typhimurium* chromosome, your mutation can be mapped more precisely by P22 transduction (Experiment 2B).

Results - Experiment 2A

Donor Strain	Auxotrophic Requirement	Recipient Strain	Auxotrophic Requirement	Map Position	#Exconjugant colonies	Comments
		MS 2	*thr*	0		
		MS 4	*pro*	7		
		MS 5	*uracil*	23		
		MS 6	*uracil*	33		
		MS 8	*his*	42		
		MS 10	*cys*	50		
		MS 12	*ser*	63		
		MS 13	*cys*	73		
		MS 100	*ile + val*	83		
		MS 14	*met*	90		

Based on these results, where does your MudJ insertion mutation map?

Based on the predicted map position, what are the most likely genes mutated?

2B. Genetic Mapping by Cotransduction

P22 transduction can be used to map more precisely the position of your MudJ mutation on the *S. typhimurium* chromosome. The linkage of your MudJ mutation to known genes will be determined by two-factor crosses. The principle is simple: if two genes can be brought into a cell on a single P22 transducing fragment, they must be less than about 1 min or 44 kb (the length of a P22 transducing fragment) apart on the chromosome. The further apart two genes are, the less frequently both of them will be present on a single transducing fragment. If the transducing fragment carries DNA that is homologous to the chromosomal DNA, occasionally the DNA can be transferred to the chromosome by homologous recombination. (Even if the transducing fragment carries some DNA that is not homologous to the recipient chromosome such as a transposon insertion, it can be inherited by homologous recombination between the flanking DNA.) However, since the probability of a cross-over between two genes increases the further apart the genes are, the frequency of inheritance of two genes by homologous recombination decreases the farther apart the genes are. Thus, the frequency of cotransduction decreases as the distance between two genes increases, so the frequency of cotransduction is a rough estimate of the distance between two genes. (It is important to note that the distance estimated from cotransduction is sometimes inaccurate. For example, there is often a significant difference between the frequency of cotransduction between two genes in reciprocal crosses.) If the apparent cotransduction frequency is very low (less than 1%), the cell may have been infected with two different P22 transducing phages. In such cases it is important to double-check your results with another gene that is closer.

It is possible to rapidly map many mutations in *S. typhimurium* using a collection of Tn*10del16del17* ("Tn10dTet") transposons located around the chromosome (Kukral et al., 1987). The locations of many of these Tn10dTet insertions have been mapped by determining their linkage to known mutations. We have grown P22 transducing phage on each of the Tn10dTet insertion strains. The Tn10dTet insertions can be transduced into any *S. typhimurium* strain by selecting for Tet^r. The Tn10dTet insertion integrates into the chromosome by homologous recombination of the flanking sequences. Thus, when the Tn10dTet insertion recombines into the chromosome the flanking wild type DNA sequences are also recombined onto the chromosome. Hence, if the Tn10dTet is located near the same gene that contains your MudJ mutation, occasionally the wild type copy of the gene will recombine into the chromosome yielding prototrophic transductants as shown below.

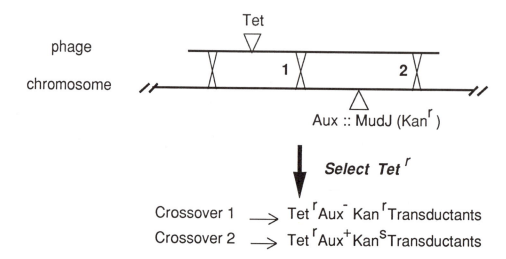

The frequency of cotransduction between the Tn10dTet and the gene containing your MudJ mutation gives a rough estimate of the distance between the two genes. Since P22 can only carry about 1 min of the chromosome, this allows you to map your mutation within a 1 min region of the *S. typhimurium* chromosome. In contrast, if the Tn10dTet insertion is not linked to your auxotrophic MudJ mutation, all of the Tet[r] transductants will still be auxotrophic as shown below.

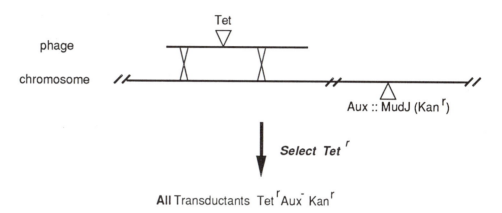

The physical distance between two point mutations can be roughly estimated from their cotransduction frequency using the Wu formula:

$$D = L \left(1 - \sqrt[3]{C}\right)$$

where:

      D = distance between two genetic markers in min or kb
      L = length of the transducing fragment (1 min or 44 kb for P22)
      C = cotransduction frequency (expressed as a decimal)

This calculation assumes that the entire length of the transduced fragment is homologous to the chromosome (i.e., there are no insertions or deletions). Since a 3.5 kb insertion mutation is present on the transducing fragment (Tn10dtet) a rough correction can be made by assuming that the size of the transducing fragment is only 41.5 kb. [A more rigorous mathematical correction described by Sanderson and Roth (1988) should be used for large insertions or cotransduction of two insertion mutations.]

References

Ingraham, J., O. Maaloe, and F. Neidhardt. 1983. *Growth of the Bacterial Cell*, pp. 409-415. Sinauer Associates, Sunderland, MA.

Kukral, A., K. Strauch, R. Maurer, and C. Miller. 1987. Genetic analysis in *Salmonella typhimurium* with a small collection of randomly spaced insertions of transposon Tn10del16del17. *J. Bacteriol. 169:* 1787-1793.

Sanderson, K., and J. Roth. 1988. Linkage map of *Salmonella typhimurium*, Edition VI. *Microbiol. Rev. 52:* 485-532.

Genetic mapping by cotransduction:

1. Grow an overnight culture of your auxotrophic MudJ insertion in 2 ml NB at 37°C.
2. Examine the genetic map of S. *typhimurium* to determine potential map positions of known mutations with the same auxotrophic requirement as your mutant (see Sanderson and Roth, 1988). In your proposal list the gene, phenotype, and the map location of all of these mutations and ask for P22 lysates on Tn10dTet insertions that map near these genes.
3. Dilute each P22 lysate 1/50 in T2 buffer, then transduce your MudJ mutant as follows:

Plate	ml cells	ml phage	
A	0.1	—	Cell control
B	—	0.1	Phage control
C	0.1	0.1	

   Spread the cells and phage directly on NB + Tet plates. Incubate overnight at 37°C.
4. Replica plate onto E + glucose and NB + Kan plates. Incubate overnight at 37°C.
5. Count the number of colonies on the NB + Tet plates and the number of Tet^r colonies that are Kan^s and grow on E + glucose plates. Calculate the cotransduction frequency as shown below:

$$\text{Cotransduction frequency} = \frac{\text{Number Tet}^r \text{ Kan}^s \text{ Aux}^+}{\text{Total Number Tet}^r}$$

If you do not get any Tet^r transductants after step 3 and you are sure you did everything correctly, there are three potential explanations: (1) the titer of the phage lysate was too low, (2) your MudJ mutant is P22^r, or (3) the strain the phage was grown on is not Tet^r. The first two possiblities are the most likely. Therefore, if you have this problem, check the titer of your phage stock (page 14) and double-check the sensitivity of your MudJ mutant to phage H5 (page 16).

Results - Experiment 2B

Donor Strain	Tn*10*		Recipient Strain	Auxotrophic Requirement	#Transductants			Comments
	min	cotransduction frequency			Tet ^{r}	Kan ^{r}	Aux $^+$	

Based on the Wu formula, what is the predicted distance between your mutation and any cotransducible Tn*10*dTet insertions?

ISOLATION OF REGULATORY MUTANTS

The purpose of this experiment is to select for regulatory mutants that increase the expression of the MudJ operon fusion isolated in Experiment 1.

SELECTION OF REGULATORY MUTANTS

In addition to indicating the relative transcription of a gene, Mud insertion mutants are also very useful for isolating regulatory mutants that overexpress the gene (Beckwith, 1981; Silhavy and Beckwith, 1985). Selection for regulatory mutants is much easier if you can distinguish high and low expression on plates, but no simple plate assays are available for many genes. This problem can be overcome using *lac* fusions to a gene. Many different media are available that distinguish high and low expression of the lac operon (see Silhavy et al., 1984). In addition, often Mud fusions that are Xgal⁺ do not express high enough Lac enzyme levels for growth on lactose as a sole carbon source, so increased *lac* expression can be directly selected by growth on lactose plates.

In this experiment, we will select for mutants that increase the expression of *lacZ* from Xgal⁺ MudJ insertion mutants. Since MudJ lacks transposase, the insertions are completely stable. If Mud vectors that are capable of transposition are used, mutants with increased *lacZ* expression are more common but most of these mutants are due to transposition of the Mud to secondary sites on the chromosome. Two types of regulatory mutations may be isolated: (1) cis-acting mutations and (2) trans-acting mutations. Cis-acting mutations usually affect a regulatory site adjacent to the gene (e.g., an operator or promoter). Trans-acting mutations usually affect a diffusible regulatory factor (e.g., a repressor protein).

Spontaneous mutations are usually rare ($<10^{-6}$), but the mutation frequency can be enhanced by treating the cells with mutagens. A wide variety of mutagens can be used, including transposons, base modifying agents, base analogs, UV radiation, and intercalating agents. Each of these different mutagens have different effects and are useful for different purposes. Two methods of isolating point mutations are used in this experiment: generalized mutagenesis with diethylsulfate (DES) and localized mutagenesis with hydroxylamine.

DES MUTAGENESIS

DES (CH_3CH_2-O-SO_2-O-CH_2CH_3) is an alkylating agent that reacts with guanine to produce ethyl guanine. Mispairing of ethyl guanine can cause G:C to A:T transition mutations. However, in addition to causing mutations directly due to mispairing, the alkylated DNA also causes mutations indirectly by inducing error prone DNA repair (Drake, 1970).

In this procedure, a liquid culture is directly treated with DES, then the mutated cells are plated on media that selects for the desired phenotype. This procedure can induce mutations anywhere on the chromosome (generalized mutagenesis). Since the cells are mutated in a liquid culture, each mutant can divide, yielding many siblings with the same mutation. Genetic and biochemical characterization of mutants can be a lot of work, so it is important to be confident that you are not wasting time repeating work on siblings with the identical mutation. In order to avoid siblings, usually only one mutant with a specific phenotype is saved from each culture.

HYDROXYLAMINE
MUTAGENESIS

In contrast to the *in vivo* mutagenesis of cells with DES, hydroxylamine (NH$_2$OH) will be used to mutagenize DNA *in vitro*. When used *in vitro*, hydroxylamine reacts with cytosine, converting it to a modified base that pairs with adenine. This has two consequences: (1) hydroxylamine only produces G:C to A:T transitions, and (2) mutations induced by hydroxylamine cannot be reverted with hydroxylamine. (If hydroxylamine is used *in vivo*, the resulting DNA damage induces error prone repair, resulting in a wide variety of mutations.) In addition, hydroxylamine gives a high ratio of mutagenic to lethal events.

Unlike many mutagens, hydroxylamine can even mutagenize DNA packaged inside of phage heads. This allows localized mutagenesis of transducing particles (Hong and Ames, 1972; Figure 3-1). A phage lysate containing transducing particles is mutagenized with hydroxylamine *in vitro*, then cells are infected selecting for a marker on a specific transducing fragment. When the transducing fragment is recombined onto the chromosome, only the small, localized region carried on that transducing fragment is mutagenized. The extent of mutagenesis can be monitored by following the mutagenesis of phage present in the lysate (Figure 3-2). Since each mutant is directly selected from an independently mutagenized transducing fragment, problems with isolation of siblings are avoided.

By using this technique it is possible to even heavily mutagenize a small region of the chromosome without mutagenizing the rest of the chromosome. Localized mutagenesis is especially useful for obtaining rare cis-dominant regulatory mutants linked to a gene (Hahn and Maloy, 1986) or rare types of mutations in a structural gene (Dila and Maloy, 1987; Myers and Maloy, 1988).

Hydroxylamine can also be used to mutate purified plasmid DNA *in vitro*. The procedure for mutagenesis of plasmids is essentially identical to the procedure for phage, except after mutagenesis the hydroxylamine is removed by dialysis prior to transforming cells (Klig et al., 1988). The transformants are selected on antibiotic plates, then the colonies are replica plated onto appropriate media to screen for the desired mutations. It is a good idea to assay mutagenesis of another plasmid gene as a control. [Some plasmids seem to be harder to mutagenize than others. This may be due to secondary structure in the DNA, which strongly inhibits mutagenesis by hydroxylamine (Drake, 1970). If this is a problem, it may be necessary to partially denature the DNA by doing the mutagenesis at a higher temperature (Humphreys et al., 1976).]

Figure 3-1. Localized mutagenesis with hydroxylamine

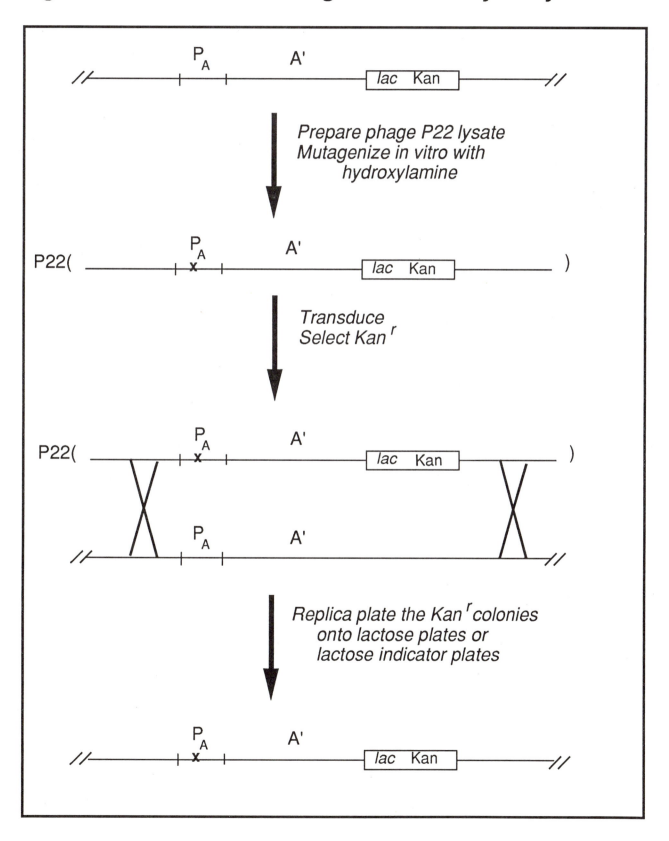

Figure 3-2. Hydroxylamine mutagenesis of phage P22

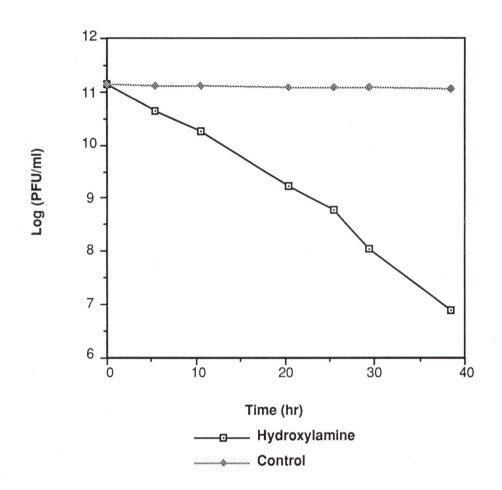

References

Beckwith, J. 1981. A genetic approach to characterizing complex promoters in *E. coli*. *Cell 23:* 307-308.

Dila, D., and S. Maloy. 1987. Proline transport in *Salmonella typhimurium: putP* permease mutants with altered substrate specificity. *J. Bacteriol. 168:* 590- 594.

Drake, J. 1970. *The Molecular Basis of Mutation*, pp. 152-155. Holden-Day, San Francisco, CA.

Hahn, D., and S. Maloy. 1986. Regulation of the put operon in *Salmonella typhimurium* characterization of promoter and operator mutations. *Genetics 114:687-703.*

Hong, J., and B. Ames. 1971. Localized mutagenesis of any specific small region of the bacterial chromosome. *Proc. Natl. Acad. Sci. USA 68:* 3158-3162.

Humphreys, G., G. Willshaw, H. Smith, and E. Anderson. 1976. Mutagenesis of plasmid DNA with hydroxylamine: Isolation of mutants of multi-copy plasmids. *Mol. Gen. Genet. 145:* 101-108.

Klig, L., D. Oxender, and C. Yanofsky. 1988. Second-site revertants of *Escherichia coli trp* repressor mutants. *Genetics 120:* 651-655.

Myers, R., and S. Maloy. 1988. Mutations of *putP* that alter the lithium sensitivity of *Salmonella typhimurium*. *Mol. Microbiol. 2:* 749-755.

Silhavy, T., and J. Beckwith. 1985. Uses of *lac* fusions for the study of biological problems. *Microbiol. Rev. 49:* 398-418.

Silhavy, T., L. Enquist, and M. Berman. 1984. *Experiments with Gene Fusions*, pp. 266-271. Cold Spring Harbor Laboratory, NY.

Smith-Keary, P. 1975. *Genetic Structure and Function*, pp. 150-152, 170-176. John Wiley and Sons, NY.

3A. Mutagenesis with Diethylsulfate (DES)

• CAUTION — Mutagens are potential carcinogens. Wear gloves and do NOT mouth-pipet. Dispose of all waste containing the mutagen in appropriate biohazard waste containers.

1. Add 50 µl of DES to a 10 ml screw-capped test tube containing 2.5 ml of E medium with no carbon source. Tighten the cap. Vortex the tube then place it in a 37°C water bath for 10 min to form a saturating solution of DES.
2. Add 50 µl of an overnight culture (about 10^9 cells per ml) to the aqueous phase. Do not shake the tube when adding the cells. Also add 50 µl of the overnight culture to a control tube containing 2.5 ml E medium without DES.
3. Incubate 50 min at 37°C without shaking.
4. Remove 50 µl and subculture into 2 ml NB.
5. Grow overnight at 37°C. This allows the treated cells to recover from the mutagenesis and mutant chromosomes to segregrate from nonmutant sister chromosomes. After overnight growth the culture will contain approximately 10^9 cells/ml.
6. Dilute the mutagenized and control cultures 10^{-6} in sterile 0.85% NaCl.
7. Plate 0.1 ml of the dilution on NB plates. Incubate at 37°C overnight.
8. Replica plate onto the following plates spread with the required auxotrophic supplement:
    E + glucose + Xgal + the required auxotrophic supplement
    NCE + lactose + the required auxotrophic supplement
9. Incubate the plates overnight at 37°C. (The NCE + lactose plates may take 2-3 days.)
10. Examine the plates. Note the number of colonies on each plate and the number of colonies on the Xgal plates that are darker blue or lighter blue than the parent strain. As a control for the efficiency of mutagenesis, also screen for new auxotrophic mutants (colonies that grow on the NB plate but not the E + glucose plate supplemented with the original auxotrophic requirement).
11. Pick several colonies that overexpress Lac and streak for single colonies on the same type of plate. Incubate overnight at 37°C.
12. Streak the purified regulatory mutants on NB plates and grow overnight at 37°C. Store the plates in the refrigerator until use.

Reference

Roth, J. 1970. Genetic techniques in studies of bacterial metabolism. *Methods Enzymol.* 17: 1-35.

Results - Experiment 3A

Strain	Auxotrophic Requirement	NB	E + Glucose	E + Glucose + Xgal + Supplement			NCE + Lactose	Comments
				Darker Blue	Lighter Blue	Total		

Number of Colonies

3B. Hydroxylamine Mutagenesis *in Vitro*

• CAUTION — Mutagens are potential carcinogens. Wear gloves and do <u>NOT</u> mouth-pipet. Dispose of all waste containing the mutagen in appropriate biohazard waste containers.

1. Grow a phage P22 stock on your MudJ auxotrophic mutant (see page 13).
2. Add the following solutions to two separate sterile test tubes:

	Mutagenesis	Control
Phosphate-EDTA buffer	0.40 ml	0.40 ml
Sterile dH_2O	0.60 ml	1.40 ml
Hydroxylamine	0.80 ml	—
1 M $MgSO_4$	0.02 ml	0.02 ml

3. Add 0.2 ml P22 phage stock grown on your MudJ mutant to each tube. It is important to use a high titer phage stock ($\geq 5 \times 10^{10}$ pfu/ml). If necessary, the phage can be concentrated as described in steps 6 and 7.
4. Incubate 24-48 hrs in a 37°C incubator.
5. Take samples at 0 time and every 4-8 hrs by diluting 10 µl into 1 ml of cold LBSE. Titer each sample on LT2 to determine the decrease in viable phage and the proportion of clear plaque mutants (see page 14). (Remember to take the decrease in phage titer into account when planning dilutions. The clear plaque mutants will only be a few percent of the total phage.)
6. Plot pfu/ml vs time on semi-log paper. Predict when killing will reach 0.1-1.0% survivors (usually 24-36 hrs — see Figure 3-2).
7. At the last time-point remove an aliquot for titering, then centrifuge the rest at 15,000 rpm for 2 hrs at 4°C in the SS34 rotor.
8. Pour off the supernatant. Overlay the phage pellet with 0.2 ml cold LBSE. Place at 4°C overnight, occasionally swirling gently to resuspend the pellet. (Do not vortex or attack the pellet with a pipet).
9. Dilute the mutagenized phage 1/10 in T2 buffer for transductions. Transduce LT2 to Kan^r with the diluted phage. Selection for Kan^r requires phenotypic expression: spread cells and phage on NB plates as shown below.

Plate	ml LT2	ml Phage	
A	0.1	—	Cell control
B	—	0.2	Phage control
C	0.1	0.05	
D	0.1	0.1	
E	0.1	0.2	

10. Incubate the plates upside-down for 4-8 hr at 37°C.
11. Replicate onto NB + Kan plates and incubate overnight at 37°C.

12. Check the plates. There should be no growth on either of the control plates. Replica plate the transduction plates on the following media to screen for regulatory mutants:

    E + glucose + Xgal + the required auxotrophic supplement
    NCE + lactose + the required auxotrophic supplement

13. Incubate the master plate and the replica plates overnight at 37°C. (The NCE + lactose plates may take 2-3 days.)

14. Examine the Xgal and lactose plates. Note the number of colonies on each plate and the number of colonies on the Xgal plates that are darker blue or lighter blue than the parent strain. Pick several colonies that overexpress Lac and streak for single colonies on the same plate. Incubate overnight at 37°C.

15. Purify the colonies on green plates and cross-streak against phage H5 (page 16). Streak the purified regulatory mutants on NB plates and grow overnight at 37°C. Store the plates in the refrigerator until use.

Reagents

LBSE

100	ml	LB
0.2	ml	0.5 M EDTA
5.85	g	NaCl

Mix. Autoclave. Store at 4°C.

Phosphate-EDTA buffer; (0.5 M KPO_4 pH 6, 5 mM EDTA) When exposed to oxygen, hydroxylamine solutions form byproducts that are toxic to cells (probably peroxides and free radicals). This nonspecific toxicity is decreased by EDTA.

6.81 g KH_2PO_4
70 ml dH_2O
Dissolve on a stirrer.
Bring to pH 6 with 1 N KOH.
Bring to 99 ml with dH_2O.
Add 1 ml of 0.5 M EDTA.
Autoclave.

Hydroxylamine/NaOH (prepare fresh)

0.175 g   Hydroxylamine (NH_2OH)
0.28 ml   4 M NaOH
Bring to 2.5 ml with sterile dH_2O.

References

Hong, J., and B. Ames. 1971. Localized mutagenesis of any specific small region of the bacterial chromosome. *Proc. Natl. Acad. Sci. USA 68*: 3158-3162.

Davis, R., D. Botstein, and J. Roth. 1980. *Advanced Bacterial Genetics*, pp. 94-97. Cold Spring Harbor Laboratory, NY.

Results - Experiment 3B

Strain	Auxotrophic Requirement	NB + Kan	E + Glucose + Xgal + Supplement			NCE + Lactose	Comments
			Darker Blue	Lighter Blue	Total		

Number of Colonies

Also include the plot of the survival of P22 (pfu/ml) vs time on semilog paper and indicate the time mutagenesis was stopped.

4

EXPRESSION OF *lac* OPERON FUSIONS

The purpose of this experiment is to measure the relative transcription of a Xgal[+] MudJ insertion mutant and regulatory mutants. Since the MudJ operon fusions isolated in Experiment 1 express the *lacZ* gene product (β-galactosidase) from the promoter of the mutated gene, transcription of the mutant gene can be quantitated by determining the β–galactosidase activity expressed in the MudJ insertion mutants (Beckwith, 1981). By determining the β-galactosidase activity expressed in regulatory mutants or under different growth conditions, the transcriptional regulation of the mutant gene can be studied.

ß-GALACTOSIDASE ASSAY

β–galactosidase can be assayed by measuring hydrolysis of the chromogenic substrate, o-nitrophenyl-β-D-galactoside (ONPG) as shown below (Miller, 1972).

ONPG
(colorless)

Galactose
(colorless)

o-Nitrophenol
(yellow)

The amount of o-nitrophenol formed can be measured by determining the absorbance at 420 nm. If excess ONPG is added, the amount of o–nitrophenol produced is proportional to the amount of β-galactosidase and the time of the reaction. The reaction is stopped by adding Na_2CO_3 which shifts the reaction mixture to pH 11. At this pH most of the o-nitrophenol is converted to the yellow colored anionic form and β-galactosidase is inactivated. The reaction can be run using whole cells that have been permeabilized to allow ONPG to enter the cytoplasm. However, since whole cells are present, the absorbance at 420 nm is the sum of the absorbance due to o-nitrophenol and light scattering due to the cells. The contribution of light scattering can be determined by measuring the absorbance at 550 nm where o-nitrophenol does not absorb. The light scattering at 420 nm is 1.75x the light scattering at 550 nm, so the absorbance of o-nitrophenol is determined by subtracting 1.75 x OD_{550} nm. The corrected absorbance is then used to calculate the activity of β-galactosidase.

Assay the β-galactosidase activity expressed by your original MudJ mutant and any regulatory mutants you isolated in Experiment 3. In addition, you may want to determine how your MudJ fusion is regulated under different growth conditions by assaying the β-galactosidase activity in cells grown in appropriate media (e.g., when starved for the auxotrophic supplement compared to when excess supplement is available). Looking up the regulatory properties of similar mutants will give you a good idea what growth conditions to try.

References

Miller, J. 1972. *Experiments in Molecular Genetics*, pp. 352-355. Cold Spring Harbor Laboratory, NY.

ß-galactosidase assay

1. Inoculate each Mud insertion mutant in 2 ml E + glucose medium with the required supplement. Grow overnight at 37°C.
2. Add 1.5 ml of the culture to a sterile microfuge tube. Centrifuge for 20 sec in the microfuge. Pour off the supernatant and resuspend the cell pellet in 1.5 ml 0.85% NaCl. Vortex until completely resuspended.
3. Subculture 0.25 ml of each cell suspension into 5 ml medium in a Klett flask. Grow at 37°C to mid-log phase (100 - 120 Klett units).
4. Add 1.5 ml to a microfuge tube. Spin 30 sec in a microfuge. Pour off the supernatant then resuspend the cell pellet in 1.5 ml of sterile 0.85% NaCl. Vortex until completely resuspended.
5. Prepare triplicate dilutions of 0.1 ml cells with 0.9 ml complete Z–buffer in test tubes. Also prepare two controls with only 1.0 ml Z-buffer. (Save the cell suspension on ice!)
6. To permeabilize cells add 1 drop of 0.1% SDS and 2 drops of chloroform from a Pasteur pipet. Vortex.
7. Place the tubes in a 30°C water bath and allow to equilibrate for about 2 min.
8. Add 0.2 ml ONPG to each tube and vortex to initiate the reaction. Return to the 30°C shaker. Note the time.
9. When a yellow color develops, stop the reaction by adding 0.5 ml of 1 M Na_2CO_3. Note the time that each reaction is stopped.
10. Return to the shaker for approximately 5 min.
11. Determine the absorbance within 1 hr. Measure OD_{420} and OD_{550} for each tube.
12. Measure OD_{650} of the cell suspension. If the OD_{650} is greater than 1.2, dilute the cells (0.5 ml cells + 0.5 ml 0.85% NaCl) and reread the absorbance. If the cells are diluted, remember to correct the OD by the dilution factor before calculating ß-galactosidase activity.
13. Using the equation shown below, calculate the ß-galactosidase activity of each sample. Determine the mean and standard deviation of the triplicates.

$$\text{Activity} = \frac{OD_{420} - (1.75\,OD_{550})}{OD_{650} \times \text{time} \times \text{vol}} \quad \times \quad \frac{1\ \text{nmol}}{0.0045\ \text{ml cm}} \times 1.7\ \text{ml}$$

Where:
 time = time of reaction in min
 vol = ml cells added to the assay tubes
 Σ_{420} o-nitrophenol=0.0045 OD_{420}/ml cm
 1.7 ml = total vol
 cuvette = 1 cm path length
 Activity = nmol / min / OD_{650} ml

Reagents

Z-buffer stock solution
        4.27 g Na_2HPO_4
        2.75 g $NaH_2PO_4 H_2O$
        0.375 g KCl
        0.125 g $MgSO_4 7H_2O$
        Adjust to pH 7.0.
        Bring to 500 ml with dH_2O. Do not autoclave. Store at 4°C.
For complete Z-buffer — Prior to daily use mix:
        50 ml Z-buffer
        0.14 ml ß-mercaptoethanol

ONPG (4 mg/ml) (Sufficient for 100 assays)
        80 mg o-nitrophenyl-ß-D-galactoside
                (o-nitrophenyl-ß-D-galactopyranoside)
        20 ml dH_2O

1 M Na_2CO_3 (Sufficient for 100 assays)
        5.3 g Na_2CO_3
        50 ml dH_2O

Results - Experiment 4

Strain	Auxotrophic Requirement	Growth Condition	Time (min)	OD_{420}	OD_{550}	OD_{650}	Cell Dilution	Corrected OD_{650}	ß-galactosidase activity per tube	ß-galactosidase activity $\overline{X}$	ß-galactosidase activity S.D.

SOUTHERN BLOTS

Southern blots can be used to determine the position of chromosomal deletion and insertion mutations with respect to the physical map of a gene. The purpose of this experiment is to determine the position of your MudJ insertion mutation relative to restriction sites on the chromosome.

PROPERTIES OF RESTRICTION ENZYMES

Restriction enzymes are endonucleases that recognize specific DNA sequences and digest double-stranded DNA by cleaving phosphodiester bonds in each strand of the duplex DNA. For practical purposes, restriction enzymes can be divided into two groups:

(1) Type I restriction enzymes recognize a specific DNA sequence, but cleave the DNA at some distance from this site. These types of restriction enzymes are not very useful for *in vitro* genetic engineering, but are an important consideration when trying to clone foreign genes (see Experiment 9). The sequence recognition, restriction endonuclease, and methyltransferase functions of Type I restriction systems are encoded by three genes. The enzymes form a multimeric complex that binds to a specific sequence but cuts the DNA nonspecifically between 100 to 1000 base pairs from the site. Type I restriction enzymes require Mg^{++}, ATP, and S–adenosylmethionine (SAM) for activity.

(2) Type II restriction systems are encoded by two genes. In contrast to Type I restriction enzymes, Type II endonucleases cut the DNA within or very near the recognition sequence and they only require Mg^{++} for activity. Essentially all Type II restriction endonucleases cleave DNA into fragments with 5'-phosphates and 3'–hydroxyls at the ends. Type II DNA–methyltransferases require SAM for activity. Type II restriction enzymes are the most useful tools for cloning DNA because they recognize and cleave the same site. Most restriction sites for Type II endonucleases are palindromes. For example, the restriction sites for two endonucleases used in this experiment are shown below:

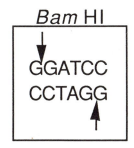

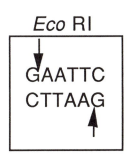

However, some restriction enzymes recognize sites with ambiguities where some of the bases are not exactly defined. Different restriction enzymes that recognize identical sequences are known as isoschizomers. Isoschizomers may have alternative cleavage sites within the recognition sequence or may show different sensitivities to methylation of the recognition sequence. Examples of some Type II restriction enzymes and their recognition sites are shown in Appendix 6A.

Several factors may affect the activity and specificity of restriction enzymes, including temperature, pH, ionic strength, and structure of the DNA. In addition, DNA fragments may be modified by the host (e.g., *hsd* or *dam* methylases): if the modification overlaps a restriction site, it may prevent cleavage of that site by a restriction enzyme. The appendix of the New England Biolabs catalog has a wealth of information on the effects of different conditions on activity of restriction enzymes. The best way to find out the optimal conditions for a specific enzyme is to look in the catalog from the supplier.

SEPARATION OF DNA FRAGMENTS

Different size DNA fragments can be separated by agarose gel electrophoresis. DNA fragments have a constant charge/length ratio due to the net negative charge of the phosphate backbone. Therefore, DNA migrates toward the (+) electrode. During electrophoresis DNA molecules seem to "snake" through the pores in the gel "head first." The rate of migration of linear double-stranded DNA is inversely proportional to the $\log_{10}$ of its molecular weight: smaller DNA fragments snake through the pores easier and hence migrate faster.

SOUTHERN BLOTS

Southern blots are used to transfer DNA fragments from an agarose gel to a membrane for hybridization. Chromosomal DNA is digested with a restriction endonuclease and the restriction fragments are separated by size on an agarose gel. Since restriction sites are spaced relatively randomly in the chromosome, the size of the restriction fragments varies from very small to very large. Large fragments do not transfer as well as small fragments so the DNA in the gel is partially depurinated with acid to break large fragments into smaller pieces. Then the DNA is denatured with NaOH. Throughout these treatments the DNA fragments remain in the original position in the agarose gel. The DNA fragments are transferred from the agarose gel to a nitrocellulose or Nylon membrane by blotting: the membrane is laid on the agarose gel and dry filter paper is laid on top of the membrane, causing the buffer and DNA to be drawn from the gel to the membrane by capillary action. Once transferred to the membrane the DNA is immobilized. The membrane is then prehybridized with salmon sperm DNA which binds nonspecifically to the membrane, preventing the radioactive probe from binding nonspecifically. Then, a radioactive single stranded DNA probe is hybridized to the membrane. If the DNA probe is complementary to the chromosomal DNA bound to the membrane, it will specifically hybridize to that band of the membrane. The membrane is washed to remove any weakly bound probe, then the radioactive bands are visualized by autoradiography (see Figure 5-1).

HYBRIDIZATION CONDITIONS

Hybridization between the probe and the chromosomal DNA requires formation of H-bonds between the complementary sequences. Association (annealing) and dissociation (denaturation) of a double helix is a highly cooperative process. The stability of double stranded DNA determines its T_m, the temperature at which 50% of the DNA is denatured. Several factors affect the T_m, including: the length of complementary base pairs, the ratio of G:C to A:T in the DNA, the number of mismatches between the two DNA strands, the ionic strength of the solution, and the concentration of formamide (see Ausubel et al., 1988). As the conditions approach the T_m, annealing between imprecise DNA hybrids is decreased. This decreases the background due to nonspecific annealing with the probe but also decreases the annealing between similar but not precise hybrids. Thus, the closer the conditions are to the T_m the greater the probability that only precise matches will anneal (high stringency). The conditions used here, hybridization at 42°C in 50% formamide, require a match of 95-100% for hybridization. The membrane is also washed thoroughly with detergent (SDS) and with high stringency conditions to remove any probe that may still be nonspecifically bound to the membrane. Sometimes you want to detect hybridization between DNA fragments that have less similarity (e.g., if you wanted to use a cloned *S. typhimurium* gene as a probe for the analogous gene from another organism). By adjusting the temperature, ionic strength, or formamide concentration the stringency of the hybridization conditions can be optimized for the desired match (Ausubel et al., 1988; Maniatis et al., 1982).

AUTORADIOGRAPHY

X-ray film is sensitive to radiation in addition to light. The film is coated with an emulsion of silver halides suspended in gelatin. When radiation strikes a silver halide crystal, the crystal absorbs energy and releases an electron. The electron is attracted to a positively charged silver ion, forming an atom of metallic silver. The film developer amplifies this effect by reducing exposed silver halide crystals to metallic silver. Finally, the film is rinsed in a fixer which converts any silver halide that was not reduced by the developer into soluble silver thiosulfate. After rinsing with water to remove the fixer and drying, the film retains the exposed silver grains over areas of exposure to radiation. [See Hahn (1983) for a thorough discussion of how to choose the type of film, the use of intensifying screens, etc.]

INTERPRETATION

The physical map of MudJ insertion mutations can be determined from the autoradiogram. A partial restriction map of MudJ is shown below.

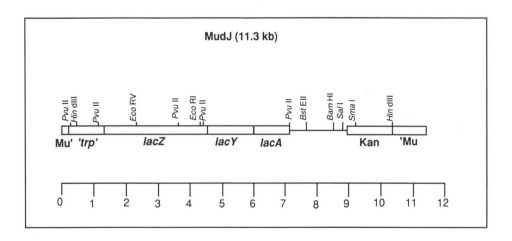

Bam H1 and *Eco* RI cut at known positions in MudJ. These restriction enzymes will also cut the chromosome at sites on either side of a MudJ insertion, but the exact distance of the cuts in the flanking DNA will vary for different MudJ insertion mutants. When the chromosome from a MudJ mutant is digested with the restriction enzyme, Southern blotted, and hybridized to a radioactive probe of MudJ DNA, any band that contains part of the MudJ insertion will hybridize. Thus, by digesting chromosomal DNA from a MudJ insertion mutant with *Bam* H1, *Eco* RI, and *Bam* H1 + *Eco* RI the distance and orientation of these sites can be determined relative to the MudJ mutation. Based on the size of the fragments, the distance and orientation of the *Bam* H1 and *Eco* RI sites can be determined relative to the MudJ insertion (Figure 5-2).

Figure 5-1. Southern blots

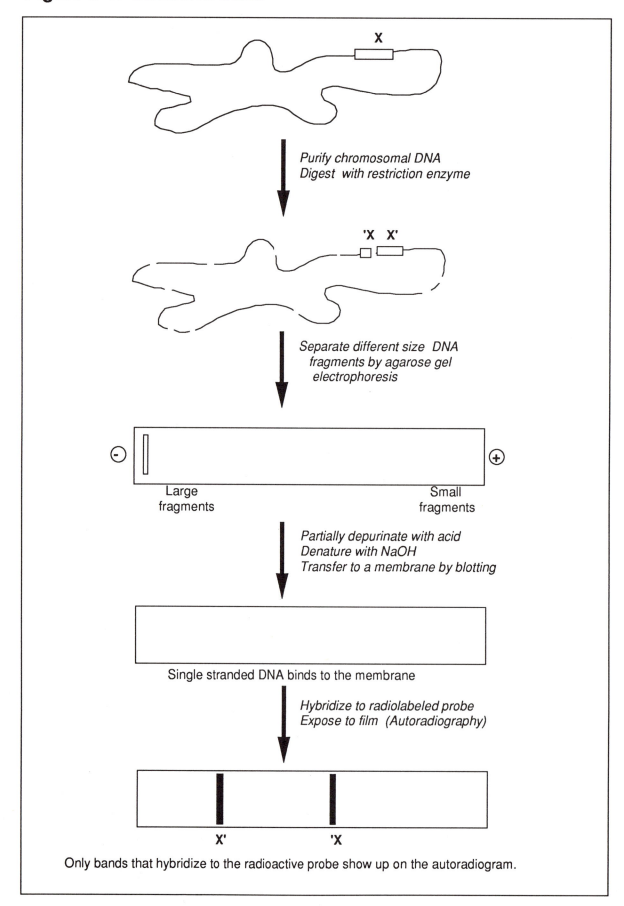

Purify chromosomal DNA
Digest with restriction enzyme

Separate different size DNA
fragments by agarose gel
electrophoresis

Large
fragments

Small
fragments

Partially depurinate with acid
Denature with NaOH
Transfer to a membrane by blotting

Single stranded DNA binds to the membrane

Hybridize to radiolabeled probe
Expose to film (Autoradiography)

X'   'X

Only bands that hybridize to the radioactive probe show up on the autoradiogram.

Figure 5-2. Mapping restriction sites near a chromosomal MudJ insertion by Southern blots

The orientation and distance of restriction sites from a chromosomal MudJ insertion mutant can be determined by Southern blots and probing with a radiolabeled MudJ probe as shown below.

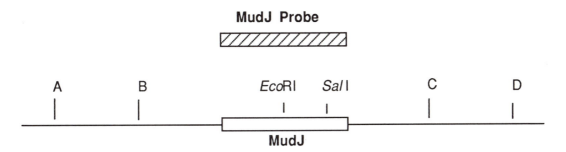

Only fragments that hybridize to the radioactive probe will show up on the autoradiogram. Thus, when digested with either *Sal* I or *Eco*RI two bands will be visible since each enzyme will cut once inside of MudJ, and only the two fragments from the closest cuts on either side of MudJ will hybridize to the probe. When digested with both enzymes three bands will be visible: a 4.2 kb *Eco*RI-*Sal* I fragment from within MudJ, and two fragments from the closest sites on either side of MudJ (i.e. B - *Eco*RI and *Sal* I - C).

For Example:

Enzyme	Size fragments (kb)
*Eco*RI	18 + 5.5
Sal I	12.7 + 9
*Eco*RI + *Sal* I	9 + 5.5 + 4.2

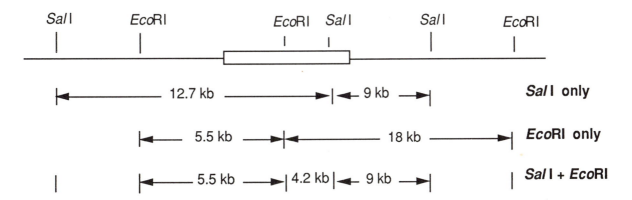

However, note that if both *Eco*RI or both *Sal* I sites are on the inside then it will not be possible to orient the outside sites without doing additional restriction digests.

NOTE. This experiment has many steps. It takes a relatively long time from the isolation of chromosomal DNA to the development of the autoradiogram. In order to keep the whole experiment in perspective it is important to constantly remind yourself why you are doing each step.

References

Ausubel, F. et al. 1988. *Current Protocols in Molecular Biology*, pp. 2.9.1-2.9.10. John Wiley and Sons, NY.

Hahn, E. 1983 (July). Autoradiography: A review of basic principles. *Am. Laboratory*.

Maniatis, T., E. Fritsch, and J. Sambrook. 1982. *Molecular Cloning*, pp. 388-389. Cold Spring Harbor Laboratory, NY.

Silhavy, T., M. Berman, and L. Enquist. 1984. *Experiments with Gene Fusions*, pp. 186-188. Cold Spring Harbor Laboratory, NY.

Southern, E. 1975. Detection of specific sequences among DNA fragments separated by gel electrophoresis. *J. Mol. Biol. 98*: 503-517.

5A. Purification of chromosomal DNA

1. Grow separate cultures of the MudJ mutant and LT2 overnight in 5 ml NB. (LT2 is a negative control for the Southern blots — chromosomal DNA from LT2 lacks substantial homologies to the probe DNA so it should not hybridize.)
2. Add 1.5 ml cells to a microfuge tube and spin for 30 sec. Pour off the supernatant, refill the tube, and spin again.
3. Resuspend the pellet in 567 µl TE. Vortex thoroughly.
4. Add 30 µl of 10% SDS and 3 µl of Proteinase K. Mix by inverting then incubate at 37°C for 1 hr.
5. Add 0.7 ml of phenol:chloroform:isoamyl alcohol (25:24:1). Mix thoroughly by inverting the tube until an emulsion is formed then spin 5 min in a microfuge.
6. Remove the supernatant with a Pasteur pipet being careful to avoid the interphase. Add the supernatant to a clean microfuge tube.
7. Add 0.6 vol of isopropanol to precipitate the DNA. Gently invert the tube until stringy white DNA fibers appear.
8. Spin for 10 min in a microfuge.
9. Pour off the supernatant and add 1 ml of ice cold 70% ethanol. Spin for 5 min in a microfuge.
10. Pour off the supernatant and drain the tube over a clean Kimwipe. Dry the DNA pellet about 15 min in a vacuum dessicator.
11. Add 100 µl TE. Resuspend the DNA by gently inverting the tube. The DNA must be completely dissolved before proceeding. Sometimes it is necessary to dissolve the DNA overnight on a rocker at 4°C.

5B. Restriction digests of chromosomal DNA

Restriction enzymes (RE) are expensive and they can be easily ruined. By following these guidelines you can avoid wasting restriction enzymes and your time:

- Most restriction enzymes are stored at -20°C in 50% glycerol. Many enzymes lose activity if allowed to warm up (even on ice). Therefore, when doing restriction digests prepare all reagents except the restriction enzyme. Then remove the enzyme from the freezer and immediately place on ice. Add the enzyme then return it to the freezer as soon as possible.

- It is important to avoid contaminating the restriction enzymes with nucleases. Fingers carry a lot of nucleases so wear gloves to prevent contamination of the enzyme when opening the microfuge tube. Use a new, sterile pipet tip every time you remove an enzyme or buffer.

- A high concentration of glycerol can alter the activity of many restriction enzymes, therefore the volume of restriction enzyme added should be less than 1/10 of the final volume of the reaction mixture.

- Often the amount of enzyme can be decreased if the digestion time is increased. Small aliquots can be removed from the reaction and run on a minigel to monitor the progress of the digestion. Sometimes a DNA sample will not cut no matter how long you give it. This is usually due to contaminants in the DNA. Many interfering contaminants can be removed by another ethanol precipitation or by drop dialysis (Appendix 5).

1. Digest chromosomal DNA from the insertion mutant with *Bam* H1, *Eco* R1, and *Bam* H1 + *Eco* R1 in 3 separate microfuge tubes. Also digest chromosomal DNA from LT2 with *Eco* R1 as a control.

dH$_2$0	___µl
DNA (up to 1 µg)	___µl
10x RE buffer	2.5µl
RE (10 U / µg DNA)	1.0µl
Total	25.0 µl

    Mix well by pipetting up and down with a pipetman. Incubate at 37°C overnight.

2. It is important that the DNA be completely digested. To check for complete digestion, mix 2 µl of the digested DNA + 8 µl TE + 1 µl Blue II. Run the digested DNA and Lambda *Hind*III standards for about 30 min on a 0.8% agarose minigel (Appendix 7A). Stain the gel with ethidium bromide and examine it on the UV transilluminator (Appendix 7D). A complete digest should give a uniform smear from the size of the highest molecular weight Lambda band down to the low molecular weight bands.

3. Add 2 µl Blue II to the remaining DNA digest. Run the digested DNA and Lambda *Hind*III standards overnight on a large 0.8% agarose gel (Appendix 7A). By loading the gel asymmetrically it will be easier to keep track of which DNA is where. Be sure to note the order of the samples loaded.

4. Stain the gel with 0.5 µg/ml ethidium bromide and photograph on the UV transilluminator (Appendix 7D).

5C. Southern blots

NOTE: Agarose gels are fragile and can be easily broken when moved from one solution to another, so always support the gel on a glass plate when moving it. WEAR GLOVES — oil from your hands can ruin the hybridization membrane.

1. Sequentially soak the gel in the following solutions. Gently rock the gel in each solution.
    a. 200 ml depurination solution for about 15 min.
    b. Rinse 2-3 times with dH$_2$O.
    c. 200 ml denaturation solution for about 30 min.
    d. 200 ml neutralization solution for about 30 min.
2. While the gel is soaking, prepare the blotting setup as follows: Cut 10 pieces of Whatman 3MM paper, 1 piece of Nylon membrane and an 8-10 cm stack of paper towels to the exact size of the gel. Cut 2 pieces of 3MM paper the width of the gel but long enough to form wicks (see the diagram below).
3. Wet the membrane in a dish of distilled water. (Many types of Nylon membranes will not wet uniformly if buffer is used instead of water. If there are any spots on the membrane that do not wet, they may prevent transfer of the DNA. The best solution is to get a new membrane.)
4. Fill a baking pan about halfway with 20x SSC. Wet the two 3MM paper wicks in the pan. Lay a glass plate sideways over the baking pan so it is supported by the sides of the pan. Lay the two 3MM paper wicks over the plate so they drape into the pan on either side of the glass plate. Stack 3 sheets of the gel-sized 3MM paper on the glass plate on top of the wick. Pour enough 20x SSC on the paper to just wet the surface. Smooth out any air bubbles by gently rolling a pipet over the surface. Place the gel on top of the stack of 3MM paper. Without stretching the gel, smooth out any air spaces between paper and gel.
5. Carefully lay the membrane on top of the gel. Try to avoid trapping air bubbles under the membrane. DNA transfer begins immediately, so do NOT move the membrane once it has touched the gel.
6. Stack the remaining 6 sheets of Whatman 3MM paper on top of the membrane. Wet with just enough 20x SSC to saturate the paper. Smooth out any air spaces between the paper sheets and the membrane.
7. Stack the precut paper towels on top of the 3MM paper. It is important that the membrane, 3MM paper, or paper towels do not hang over the gel because they will bypass the movement of fluid through the gel.
8. Lay a glass plate on top of the stack. Place a plastic bottle containing about 500 ml of water on top of the glass plate. Make sure that the baking pan is about half filled with 20x SSC. Cover the open sides of the pan with plastic wrap to decrease evaporation of the SSC.
9. Leave at room temp overnight to "blot." The rate of transfer of DNA to the membrane depends on the size of DNA fragment. Small fragments (about 1 kb) transfer within 1-2 hrs but large fragments (over 15 kb) require much longer.
10. Remove the weight, towels, and 3MM paper above the gel. Carefully remove the membrane. Soak the membrane in 2x SSC for 5 min.
11. Lay the membrane on a clean piece of 3MM paper and allow it to air-dry. If necessary store the dried membrane between two sheets of 3MM paper in a sealed zip-lock bag at room temperature until use.
12. Restain the agarose gel with ethidium bromide and examine on the transilluminator to check for efficient DNA transfer. Very little DNA should remain in the agarose gel.

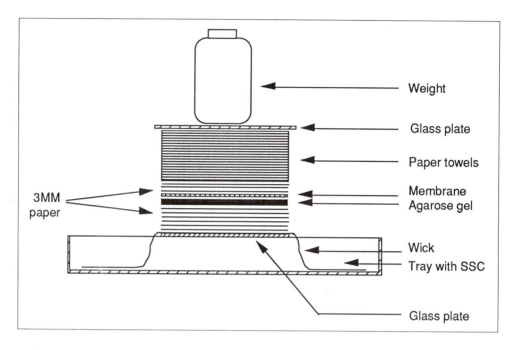

Reagents

20x SSC (3 M NaCl, 0.3 M sodium citrate)
     175.3 g NaCl
      88.2 g Na$_3$Citrate 2H$_2$0
     Dissolve in 800 ml ddH$_2$0.
     Adjust to pH 7.0 with HCl.
     Bring to 1000 ml with ddH$_2$0.

       Add 20 ml of 20x SSC to 180 ml ddH$_2$O for 2x SSC.

Depurination solution (0.2 M HCl)
     10 ml 6 N HCl
   290 ml ddH$_2$O

Denaturation solution (0.5 M NaOH, 1.5 M NaCl)
     20 g NaOH
     88 g NaCl
     Dissolve in about 800 ml dH$_2$O.
     Bring to 1000 ml with dH$_2$O.

Neutralization solution (1 M TrisHCl, 1.5 M NaCl)
     13.4 g Tris base
   140.4 g TrisHCl
     88   g NaCl
     Dissolve in 800 ml dH$_2$O.
     Bring to 1000 ml with dH$_2$O.

Hybridization Membrane
The original Southern procedure used nitrocellulose membranes but Nylon membranes give similar results and are easier to use. The many brands of Nylon-66 membranes we have tried seem to work equally well. However, minor modifications of the protocol may be necessary for some types of Nylon membranes. Always check the recommendations of the manufacturer before using the membranes.

5D. Nick translation of plasmid DNA

This protocol uses limited DNase I digestion to introduce nicks into the DNA. The 5' to 3' exonuclease activity of *E. coli* DNA polymerase I removes nucleotides in front of the nick while the 5' to 3' polymerase activity fills in nucleotides behind, incorporating a radioactive dNTP into the DNA.

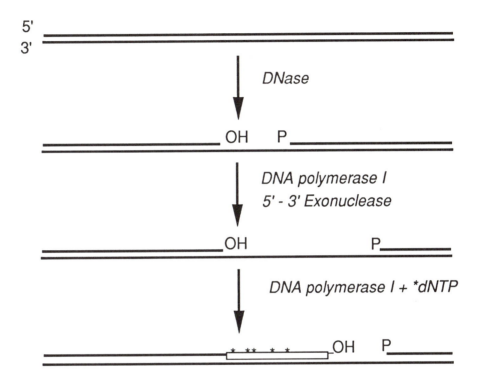

Nick translation is a quick and easy way to uniformly label DNA with radioactivity to a high specific activity. Either a whole plasmid can be labeled or, if necessary, a small DNA fragment can be purified and labeled (e.g., if the vector sequences hybridize with the nucleic acid to be probed). In this experiment a plasmid carrying Mud but no other chromosomal homology is nick translated. At the same time Lambda *Hind*III fragments are nick translated to probe the standards so they are visible on the autoradiogram.

The only tricky part of the protocol is optimizing the extent of DNaseI digestion. If there is insufficient nicking by DNase, very little radioactivity will be incorporated. On the other hand, if there is too much DNase digestion the DNA may be extensively degraded. Optimal DNase digestion conditions can be determined running several reactions with varying dilutions of DNase I. The amount of radioactivity incorporated in each reaction can be determined by precipitating a small aliquot with TCA and counting the precipitated DNA in a liquid scintillation counter. Usually the maximal amount of label incorporated is 30-40%.

NOTE: Several companies sell nick translation kits that are almost foolproof and eliminate the time and effort necessary to optimize the reagents. In addition, nonradioactive nucleic acid labeling kits are available, making it possible to avoid using radioactivity for this experiment.

CAUTION: The following experiment involves radioactivity. Wear gloves and a labcoat. Keep all radioactivity in the designated area.

*1.   Add to a microfuge tube:

Plasmid DNA (1.0 µg)	__ µl
ddH$_2$O	__ µl
Lambda HindIII standards (0.1 µg diluted in ddH$_2$O)	__ µl
dNTP solution	2.5 µl
10x Nick translation buffer	2.5 µl
Diluted DNase I	1.0 µl
DNA polymerase I holoenzyme (2.5 units)	1.0 µl
[^{35}S]dATP (high specific activity, aqueous)	2.0 µl
Total	25.0 µl

*2.   Mix with a pipetman. Incubate 1 hr in an ice-water bath at about 15°C.

*3.   Remove the unincorporated [^{35}S]dATP on a Sephadex G-50 gel filtration column:

    a. Place a microfuge tube under the column to collect the effluent. Add the entire sample to the column with a Pipetman and wait until it runs into the column.

    b. Add 200 µl TE and allow it to run into the column.

    c. Gently add 1 ml TE to the column. Collect 4 drop (about 200 µl) fractions in microfuge tubes.

    d. Determine the amount of radioactivity in each tube with a Geiger counter. Almost all of the DNA should elute as a sharp peak with the void volume. Unincorporated dNTPs should remain on the column. Dispose of the column in the radioactive waste.

*4.   In order to determine the amount of radioactivity incorporated into the probe, TCA precipitate the DNA:

    a. Spot 1 µl of the nick translated DNA onto a small filter paper disk then place it in a small scintillation vial.

    b. Add 2 ml of 10% TCA. Cap the tube. Invert to mix then leave for 5 min at room temperature.

    c. Pour off the TCA into the radioactive waste.

    d. Repeat steps b and c with 5% TCA.

    e. Repeat steps b and c with 95% ethanol.

    f. Allow to air dry thoroughly.

    g. Also spot 1 µl of the sample onto another filter disk, place it in a scintillation vial and allow it to air dry. (This is a control to determine the total radioactivity in the sample: do not TCA precipitate it.)

    h. Add scintillation fluid and count in a liquid scintillation counter. Note the cpm incorporated into DNA, the total cpm, and the percent incorporation [(cpm in DNA) x (total cpm) x 100%].

*5.   Check the work area with a Geiger counter.

NOTE. Other methods that are commonly used to remove unincorporated nucleotides include: three sequential ethanol precipitations in ammonium acetate (Appendix 5B), and "spin-columns," small gel filtration columns that are centrifuged to speed the elution (Maniatis et al., 1982; Silhavy et al., 1984). In addition, a wide variety of columns with various gel filtration or ion-exchange resins that can be used to separate large DNA fragments from unincorporated nucleotides are available commercially.

Reagents

10x Nick translation buffer

        50 µl 1 M TrisHCl pH 7.4

        10 µl 1 M $MgCl_2$

         1 µl 1 M DTT

         1 µl 100 mg/ml BSA (Pentax Fraction V)

        38 µl Sterile ddH_2O

dNTP solution

        10 µl 200 µM dCTP

        10 µl 200 µM dGTP

        10 µl 200 µM dTTP

DNase I stock solution (1 mg/ml)

        5.0 mg DNase I (RNase-free, 2000-2500 units/mg)

        2.5 ml Sterile glycerol

        0.5 ml 1 M TrisHCl pH 7.4

        2.0 ml Sterile ddH_2O

        Dissolve the DNase.

        Store in 100 µl aliquots at -20°C.

        Thaw on ice for about 1 hr before making the working solution.

DNase I diluted working solution

        10 ml ddH_2O

        100 µl 1 M TrisHCl pH 7.4

        50 µl 1 M $MgCl_2$

         1 µl 1 mg/ml DNase I

        Dilute immediately prior to use.

10% TCA (Trichloroacetic acid)

        Dissolve 10 g TCA in about 80 ml ddH_2O.

        Bring to 100 ml with ddH2O.

Sephadex G-50 columns:

    Hydrate the resin by adding 10 g Sephadex G-50 (medium) to 100 ml TE and autoclaving for 15 min.

    Allow to cool at room temperature.

    Swirl gently and pour off most of the TE and fine Sephadex particles that do not sediment.

    Add fresh TE and store at 4°C until use.

    Use small (6 ml) fritted bottom dispo-columns or Pasteur pipets plugged with a small amount of glass wool. Swirl the flask of Sephadex to form a uniform suspension and fill the column with the suspension. Allow the Sephadex to settle then add more until the column contains about 4 ml of packed Sephadex. Packing the column takes about 5-10 min. The Sephadex columns can be stored at 4°C.

5E. DNA hybridization

CAUTION: This experiment uses radioactivity and it requires some tricky manipulations. Make sure the work area is covered with disposable absorbent paper, wear gloves and a lab coat, be careful to avoid spills, and leave a Geiger counter "on" nearby so you can frequently monitor your gloves and the work area.

1. Place the Nylon-66 membrane into a heat-sealable plastic bag. Seal 3 sides of the bag.
2. Add 10 ml hybridization solution without the radioactive probe for prehybridization.
3. Squeeze out all bubbles and seal the fourth side. Leave enough room on one side to cut and reseal the bag.
4. Place at 42°C for at least 2 hrs (can go overnight if desired).
*5. Add at least 10^6 cpm of the nick translated plasmid DNA ($>10^7$ cpm/μg DNA) to a small microfuge tube. Place in a 95°C heating block for 5 min to denature the DNA. Quickly cool the tube on ice.
*6. Add the denatured probe to 10 ml hybridization solution in a screw capped disposable polypropylene tube. Mix well.
7. Cut a corner from the bag containing the prehybridization solution. Place the bag on a flat surface and gently squeeze out the liquid by rolling a pipet over its surface. Do not let the membrane dry out!
*8. Carefully pour the hybridization solution into the bag. Carefully squeeze out any air bubbles then seal the corner of the bag.
*9. Seal the first bag in another heat-sealable bag (just in case the inner bag leaks).
*10. Place at 42°C for 24-48 hrs.
*11. Remove the outer bag. Cut open the top of the inner bag and carefully slide the membrane out. Discard the bag in the radioactive waste.
*12. Transfer the radioactive membrane sequentially to the following wash solutions. Gently rock the pan during each step. Pour the used wash solutions in the radioactive waste.
    a. 1x SSC + 0.1% SDS for 5 min at room temperature.
    b. Repeat for 15 min at room temperature.
    c. 0.5x SSC + 0.1% SDS for 15 min at room temperature.
    d. 0.1x SSC + 0.1% SDS for 15 min at room temperature.
    e. 0.1x SSC + 1.0% SDS for 30 min at 42°C. Follow the elution of background radioactivity with a Geiger counter. If the background on the membrane is still high, increase the wash times.
*13. Drain the membrane, lay it on a piece of 3MM paper and dry it in a 65°C oven.

Reagents

20x SSC

        175.3 g NaCl
         88.2 g Na_3Citrate $2H_2O$
        Dissolve in 800 ml ddH_2O
        Adjust to pH 7.0 with HCl
        Bring to 1000 ml with ddH_2O.

100x Denhardt's solution

    0.5 g Ficoll (M_r 400,000)
    0.5 g Polyvinyl pyrrolidone (M_r 360,000)
    0.5 g BSA (Bovine serum albumin, Pentax fraction V)
    Add to 25 ml of 2x SSC. Stir to dissolve.
    Aliquot and store frozen at -20°C.

Salmon sperm DNA (5 mg/ml)

    Dissolve 50 mg DNA in 10 ml sterile ddH_2O in a sterile test tube (Type-III DNA sodium salt from salmon testes, Sigma Chemical Co).
    Vortex vigorously to dissolve.
    Pass several times through an 18 gauge hypodermic needle to shear.
    Phenol extract 2x with an equal volume of phenol:chloroform:isoamyl alcohol (25:24:1).
    Denature in a boiling water bath for 15 min.
    Quick cool on ice. Store 0.5 ml aliquots at -20°C.
    Determine the fragment size on a 1.2% agarose gel (Appendix 7A). An average fragment size of about 700 bp is ideal.

Hybridization solution

    4.0 ml 20x SSC
    1.0 ml 100x Denhardt's solution
    1.0 ml 10% SDS
    0.4 ml 5 mg/ml salmon sperm DNA
    10.0 ml Formamide (final concentration = 50%)
    3.6 ml ddH_2O
    Mix and filter through a 0.2 μm millipore filter.

5F. Autoradiography

*1. Make sure the membrane is completely dry. (If it is wet it will stick to the film and ruin the emulsion.) Place the 3MM paper and membrane in an X-ray film holder. In the darkroom with the lights out and the safelight on, remove a sheet of Kodak X-Omat AR film. Immediately put any unused film away and cover the film box. Lay a sheet of film on top of the membrane then close the film holder making sure that it is light tight. Once all the film is put away the lights can be turned back on.

*2. Expose the film at room temperature. The time required to expose the film depends upon the amount of probe that hybridized and the specific activity of the probe. Usually 24-48 hrs is sufficient for blots with about 1000-5000 cpm/ band.

*3. Remove the film in the darkroom with the lights out and the safelight on. Develop the film as follows:
   a. Submerge in Kodak X-ray developer and agitate intermittently for about 2 min.
   b. Rinse in water for about 30 sec.
   c. Fix for 5 min in Kodak rapid fixer.
   d. Rinse in running water for at least 15 min.
   e. Allow to air dry.

The membrane can be reexposed to film if necessary. Adjust the exposure time based on the initial intensity and the half-life of the radioisotope. (The half-life of ^{35}S is 87.4 days so the amount of ^{35}S decay is not likely to be significant for exposures less than a couple of weeks. However, ^{32}P has a half-life of 14.3 days, so when the probe is labeled with ^{32}P it is important to take the decay into account.)

RESULTS
EXPERIMENT 5

1. Include a photocopy of the agarose gel of your restriction digests.
2. Indicate the CPM incorporated into the DNA, the total CPM, and the calculated efficiency of incorporation of ^{35}S into the nick translated probe.
3. Include the autoradiogram from the Southern blot.
4. Include a plot of the size of Lambda HindIII standards vs $\log_{10}$ of the distance migrated.
5. Include a drawing of the predicted map of EcoRI and SalI sites near your MudJ insertion including the sizes of the restriction fragments.

6

ISOLATION OF COMPLEMENTING CLONES

The purpose of this experiment is to isolate a plasmid clone that complements the MudJ insertion mutant.

THE DNA LIBRARY

The library we will use contains a partial *Sau* 3a digest of *S. typhimurium* DNA cloned into the *Bam* H1 site of pBR328 (Hmiel et al., 1986) as shown below:

S. typhimurium DNA Library

Vector: pBR328 cut in the Tet^r gene with *Bam* HI

Inserts: 8-12 kb fragments from a *Sau* 3A partial digest of *S. typhimurium* DNA

Host: *S. typhimurium his-6165 ilv-452 metA22 trpB2 galE496 xyl-404 rpsL120 flaA66 hsdL6 hsdSA29*

Sau 3A recognizes a 4 bp restriction site that is very common in the *S. typhimurium* chromosome (on the average there is one *Sau* 3A site per 256 bp). Partial digestion with *Sau* 3A produces a nearly random population of restriction fragments. By isolating 8-12 kb fragments from a partial digest, even if there are several *Sau* 3A sites in a gene, the library is likely to carry the intact gene.

ISOLATION OF COMPLEMENTING CLONES

There are many different ways of isolating a cloned gene from a library, but the simplest way is by directly screening for complementation of a mutant defective in the desired gene. Selection for a complementing gene from a plasmid library requires: (1) the mutant gene must be recessive to the wild type allele, (2) multiple copies of the gene must not be toxic to the cell, and (3) the library must carry the intact gene. In addition, when cloning foreign genes the gene product must be expressed and functional in the host.

The plasmids can be transformed into the mutant host. Plasmids can also be transduced by P22, and P22 transduction is much more efficient than transformation. Therefore, P22 grown on the plasmid library can be used to directly transduce plasmids into *S. typhimurium* by selecting for the antibiotic resistance of the plasmid. The complementing clones can either be obtained by screening the antibiotic resistant transductants for complementing clones or by simultaneously selecting for both complementation and antibiotic resistance.

TRANSDUCTION OF PLASMIDS BY P22

The mechanism of plasmid transduction is diagrammed in Figure 6. Normally plasmids replicate by Oreplication because RecBCD nuclease quickly degrades any free ends generated by rolling circle replication. However, when P22 infects cells the P22 Abc protein ("anti-RecBCD") is expressed which inhibits RecBCD allowing plasmids to undergo rolling circle replication (Poteete et al., 1988). Rolling circle replication of the plasmids generates long concatemers of plasmid DNA. Phage P22 can package plasmid concatemers >48 kb into transducing particles (see Sanderson and Roth, 1983). Upon injection into a new recipient the plasmid DNA circularizes by recombination between plasmid repeats. Due to the high copy number of the plasmids used and the multiple copies on each concatemer, the frequency of plasmid transduction is very high (usually at least 10 times greater than for a chromosomal gene).

P22 INFECTION OF *galE* MUTANTS

S. typhimurium galE mutants can be transformed more efficiently than wild type strains (Lee and Ames, 1984), an advantage when preparing a genomic library. The *S. typhimurium* host for the plasmid library contains a *galE* mutation, so a short explanation of galactose metabolism and how P22 adsorbs to cells is necessary. P22 adsorbs to *S. typhimurium* by binding to a specific receptor on the outer membrane. The P22 receptor is the terminal portion of the lipopolysaccharide (LPS), called the O-antigen. In mutants that do not make the terminal portion of the LPS, P22 cannot adsorb to the cell so the cell is P22^r. In contrast, the receptor for phage P1 is part of the LPS core. When the complete LPS is made, the terminal portion of the LPS blocks the P1 receptor so the cell is P1^r. However, in mutants that do not make the terminal portion of the LPS, the P1 receptor is accessible so cells are P1^s(Ornellas and Stocker, 1974).

UDP-galactose is required for synthesis of the complete LPS. UDP-galactose can be made by two pathways (Nikaido, 1961). When exogenous galactose is available it can be converted to UDP-galactose by galactokinase (*galK*) and galactose-1-phosphate uridyl transferase (*galT*). When no exogenous galactose is available, UDP-galactose can be made from glucose-1-phosphate by the enzymes UDP-glucose pyrophosphorylase (*galU*) and UDP-glucose epimerase (*galE*).

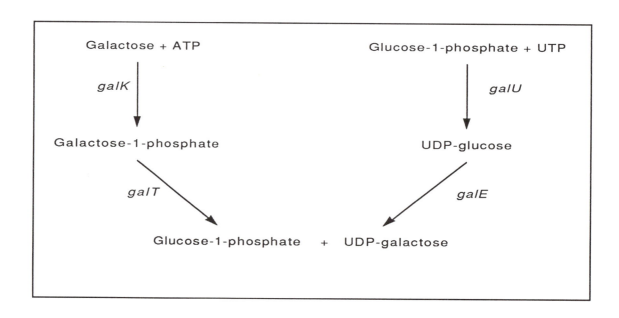

Figure 6. Transduction of plasmids by P22

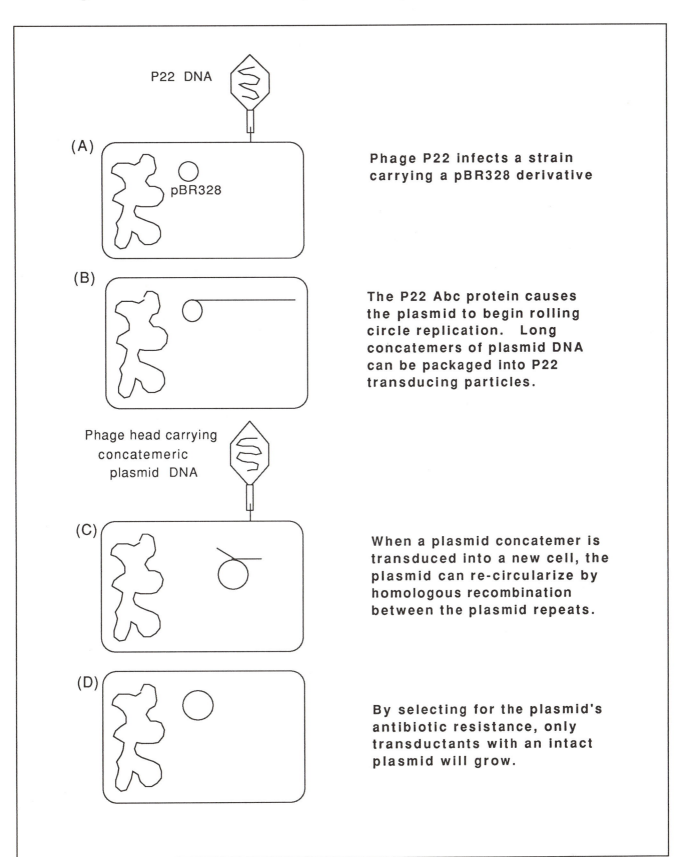

P22 DNA

(A) Phage P22 infects a strain carrying a pBR328 derivative

pBR328

(B) The P22 Abc protein causes the plasmid to begin rolling circle replication. Long concatemers of plasmid DNA can be packaged into P22 transducing particles.

Phage head carrying concatemeric plasmid DNA

(C) When a plasmid concatemer is transduced into a new cell, the plasmid can re-circularize by homologous recombination between the plasmid repeats.

(D) By selecting for the plasmid's antibiotic resistance, only transductants with an intact plasmid will grow.

Synthesis of the complete LPS can be easily controlled in *galE* mutants. When no exogenous galactose is added, galE mutants make a truncated LPS and are $P22^r$ $P1^s$. In contrast, when exogenous galactose is added, *galE* mutants make the complete LPS and are $P22^s$ $P1^r$. The *galE* mutants do not grow well in media containing 0.2% galactose because they accumulate toxic levels of galactose-1-phosphate (Fukasawa and Nikaido, 1961). However, *galE* mutants grow normally in media with low concentrations of galactose or glucose + galactose because the *gal* enzymes are not fully induced (deCrombrugghe and Pastan, 1980): there is sufficient *gal* expression for UDP-galactose synthesis but not enough to cause excess accumulation of galactose-1-phosphate.

References

de Crombrugghe, B., and I. Pastan. 1980. Cyclic AMP, the cyclic AMP receptor protein and their dual control of the galactose operon. *In* J. Miller and W. Reznikoff (eds.), *The Operon*, pp. 303-324. Cold Spring Harbor Laboratory, N.Y.

Fukasawa, T., and H. Nikaido. 1961. Galactose sensitive mutants of *Salmonella*. II. Bacteriolysis induced by galactose. *Biochim. Biophys. Acta 48*:470-483.

Hmiel, S., M. Snavely, C. Miller, and M. Maguire. 1986. Magnesium transport in *Salmonella typhimurium*: characterization of magnesium influx and cloning of a transport gene. *J. Bacteriol. 168*: 1444-1450.

Lee, G., and G. Ferro-Luzzi Ames. 1984. Analysis of promoter mutations in the histidine transport operon of *Salmonella typhimurium*: use of hybrid M13 bacteriophages for cloning, transformation, and sequencing. *J. Bacteriol. 159*:1000-1005.

Nikaido, H. 1961. Galactose sensitive mutants of *Salmonella*. I. Metabolism of galactose. *Biochim. Biophys. Acta 48*:460-469.

Ornellas, E., and B. A. D. Stocker. 1974. Relation of lipopolysaccharide character to P1 sensitivity in *Salmonella typhimurium*. *Virology 60*:491-502.

Poteete, A., A. Fenton, and K. Murphy. 1988. Modulation of *Escherichia coli* RecBCD activity by the bacteriophage Lambda Gam and P22 Abc functions. *J. Bacteriol. 170*: 2012-2021.

Sanderson, K., and J. Roth. 1983. Linkage map of *Salmonella typhimurium*, edition VI. *Microbiol. Rev. 47*: 410-453.

1. Grow the library pools overnight in LB + 0.05% galactose + 20 µg/ml chloramphenicol (Cam).
2. Grow phage P22 on the DNA library as described on page 20.
3. To select for complementing clones, spread the recipient strain and P22 grown on the DNA library on NB plates as follows:

Plate	ml cells	ml P22	
A	0.1	—	Cell control
B	—	0.2	Phage control
C	0.1	0.2	
D	0.1	0.2	
E	0.1	0.2	

4. Incubate at 37°C for 3-5 hrs to allow phenotypic expression, then replicate onto E + glucose + 10 µg/ml Cam + 125 µg/ml Kan + EGTA plates. Incubate overnight at 37°C.
5. Streak out the transductants on green plates + 20 µg/ml Cam + 50 µg/ml Kan. Incubate overnight at 37°C.
6. Pick the light-colored colonies and cross-streak against phage H5 (page 25). Incubate overnight at 37°C.
7. Streak out the phage sensitive colonies on a E + glucose + 10 µg/ml Cam + 125 µg/ml Kan plate to recheck their phenotype. Incubate overnight at 37°C.
8. Assign a strain number to each strain that checks out. Grow the strains overnight in 2 ml LB + 20 µg/ml Cam.
9. Freeze the cultures in DMSO at -70°C (see "Freezer vials" in Appendix 2).

RESULTS EXPERIMENT 6

1. Describe the strain used and what mutation it contained.
2. Indicate the number of colonies obtained on each selection plate.
3. Indicate number of P22^s colonies isolated with the expected phenotype.
4. Describe any unusual observations or problems.

ISOLATION AND RESTRICTION MAPPING PLASMID DNA

The purpose of this experiment is to determine the restriction map of the cloned DNA. The plasmid DNA must be purified for restriction mapping. The plasmid isolation procedure removes chromosomal DNA and proteins that could interfere with restriction digests or cloning experiments.

SMALL SCALE PLASMID PURIFICATION

The plasmid containing strain is grown to late stationary phase in "plasmid broth." Plasmid broth is a very rich medium that allows cells to grow to a very high density. In addition, the high concentration of yeast extract promotes the amplification of plasmids yielding a large number of plasmids per cell. The cells are treated with EDTA which chelates divalent cations, weakening the outer membrane. (Most procedures include lysozyme to degrade the cell wall peptidoglycan, but the yield of DNA is just as high without lysozyme.) The cells are then lysed with a solution of SDS and NaOH. In addition to disrupting the cytoplasmic membrane, the SDS denatures much of the cell protein. The NaOH causes dsDNA to denature because of the high pH. The denatured ssDNA can reanneal when the pH is lowered by addition of ammonium acetate. However, the rate of reassociation is proportional to the length of the DNA. Since the plasmid DNA is small and intertwined, it reanneals much faster than the chromosomal DNA. When the mixture is centrifuged, the reannealed plasmid DNA remains in solution but the large aggregates of denatured chromosomal DNA, RNA, and protein are pelleted. The plasmid DNA is then precipitated with isopropanol. The purified DNA is usually stored in a Tris buffer containing 1 mM EDTA (TE). The EDTA chelates heavy metals that could cause nicking of the DNA and divalent cations that are required by any ("God forbid") contaminating nucleases. The DNA can be stored at -20°C or at 4°C. (It was previously thought that storage at -20°C induces nicks in DNA but recent studies by BRL suggest that even after many freeze-thaw cycles the number of nicks accumulated in supercoiled DNA were negligible. The main advantage to storing DNA at 4°C is that you do not have to wait for it to thaw.)

LARGE SCALE PLASMID PURIFICATION

The DNA from a small scale plasmid purification is sufficient for several restriction digests or cloning. However, if a lot of plasmid DNA is needed (e.g., for purification of small restriction fragments) or if very pure DNA is needed, a large scale plasmid preparation may be necessary. The initial steps of the large scale plasmid preparation are similar to the small scale preparation described above. However, after the DNA is concentrated by precipitation with alcohol, it is purified by CsCl density gradient centrifugation in the presence of ethidium bromide. Covalently closed circular plasmid DNA can be purified from any contaminating chromosomal DNA, nicked plasmid DNA, and RNA in a CsCl-ethidium bromide density gradient. When centrifuged to equlibrium, CsCl forms a density gradient and any macromolecules in the CsCl gradient will band at the equivalent density of CsCl. To separate the plasmid DNA from contaminating chromosomal DNA, ethidium bromide is added to the CsCl gradient. Ethidium bromide binds to DNA by intercalating between the bases and this causes the DNA to unwind. Since co-valently closed circular plasmid DNA has no free ends, it can only unwind a limited amount and thus only binds a limited amount of ethidium bromide. In contrast, linear DNA and nicked circles don't have these topological constraints so they bind much more ethidium bromide. Ethidium bromide is less dense than DNA, therefore the density of DNA decreases as more ethidium bromide is bound. Linear and nicked circular DNA have a lower density than closed circular DNA. RNA is denser than DNA and pellets in a CsCl gradient. Thus, the lower DNA band is removed from the CsCl gradient. The ethidium bromide is extracted with butanol, the DNA is dialyzed to remove the CsCl, then the purified DNA is concentrated by ethanol precipitation.

RESTRICTION MAPPING

Once the plasmid DNA has been purified, the physical map of the cloned fragment can be determined by restriction mapping. Although many different strategies for restriction mapping are available, double digestion of DNA fragments is the most commonly used method. The DNA is completely digested with one restriction enzyme at a time and then with both restriction enzymes together. The restriction map can then be constructed by comparing the results of these digests. A restriction enzyme that cuts a circular plasmid N times will yield N DNA fragments while an enzyme that cuts a linear DNA fragment N times will yield N + 1 fragments. When cut with a second restriction enzyme, the number of fragments should be equal to the sum of the fragments from each enzyme. Any fragment that is cut by a second enzyme will result in smaller bands that add up to the same size as the uncut fragment. An example is shown in Figure 7-1 and a restriction map of pBR328 is shown in Figure 7-2.

References

Birnholm, H., and J. Doly. 1979. A rapid alkaline extraction procedure for screening recombinant plasmid DNA. *Nucleic Acids Res. 7*: 1513-1523.

Maniatis, T., E. Fritsch, and J. Sambrook. 1982. *Molecular Cloning: A Laboratory Manual*, pp. 368-369. Cold Spring Harbor Laboratory, NY.

Morelle, G. 1989. A plasmid extraction procedure on a miniprep scale. *BRL Focus* 11:7-8.

Perbal, B. 1988. *A Practical Guide to Molecular Cloning*, pp. 327-339. Wiley-Interscience, NY.

Figure 7-1. An example of restriction mapping by double digests

Purified plasmid DNA is digested with three restriction enzymes : EcoRl, HindIII, and PstI

The results are shown in the following Table:

Enzyme	Fragment sizes (kb)
EcoRl	20
HindIII	20
EcoRl + HindIII	13.5 + 6.5
PstI	10 + 7 + 3
EcoRl + PstI	9 + 7 + 3 + 1
HindIII + PstI	10 + 4.5 + 3 + 2.5

Both EcoRl (a) and HindIII (b) only cut the plasmid once .

The EcoRl site is 6.5 kb from the HindIII site (c). Thus, the following restriction map could be drawn:

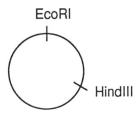

PstI cuts the plasmid at three sites.
EcoRl cuts the 10 kb PstI fragment into 9 kb + 1 kb fragments.
Based on this data four restriction maps are possible:

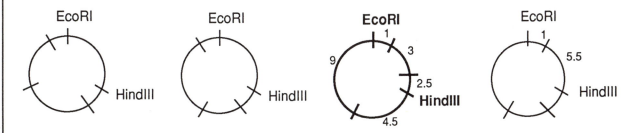

HindIII cuts the 9 kb PstI fragment into 4.5 kb and 2.5 kb fragments.
Since HindIII does not cut the 10 kb PstI fragment, the first two maps are ruled out.
Since the HindIII site is 6.5 kb from the EcoRl site, only the third map would give the size fragments obtained.

Figure 7-2. Restriction map of pBR328

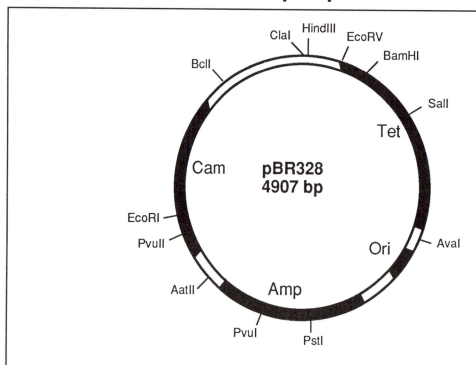

Cleavage sites of some common restriction enzymes that only cut pBR328 once:
  ClaI 24 bp, HindIII 29 bp, EcoRV 187 bp, BamHI 375 bp, SalI 651 bp, AvaI 1425 bp,
  PstI 2524 bp, PvuI 2649 bp, AatII 3201 bp, PvuII 3590 bp, EcoRI 3690 bp,
  BclI 4394 bp

Some restriction enzymes that do not cleave pBR328:
  AflII, AflIII, ApaI, BglII, HpaI, KpnI, MluI, SacI, SmaI, XbaI, XhoI, XmaI

References

Boehringer-Mannheim. *Biochemicals for Molecular Biology* Catalog, 1988.

Covarrubias, L., L. Cervantes, A. Covarrubias, X. Soberon, I. Vichido, A. Glanco, Y. Kupersztoch-Portnoy, and F. Bolivar. 1981. Construction of new cloning vehicles. V. Mobilization and coding properties of pBR322 and several deletion derivatives including pBR327 and pBR328. *Gene 13*: 25-35.

Soberon, X., L. Covarrubias, and F. Bolivar. 1980. Construction and characterization of new cloning vehicles. IV. Deletion derivatives of pBR322 and pBR325. *Gene 9*: 287-305.

7A. Small Scale Plasmid Isolation (Minipreps)

1. Grow bacteria overnight at 37°C in 2 ml plasmid broth + 20 µg/ml Cam.
2. Centrifuge the cells for 1 min in a microfuge.
3. Resuspend the pellet in 200 µl lysis solution. Leave at room temperature for 5 min.
4. Add 400 µl SDS/NaOH and mix by inverting 3-6 times. Leave on ice 5 min. (The solution will become clear and viscous.)
5. Add 300 µl 7.5 M ammonium acetate solution and gently mix for a few seconds. Leave on ice for 10 min.
6. Centrifuge for 3 min in a microfuge at 10,000 rpm.
7. Pour the supernatent into a clean microfuge tube. Be careful to avoid the loose, viscous pellet.
8. Add 0.6 vol of isopropanol (400-500 µl) and leave at room temperature for 10 min to precipitate the DNA.
9. Centrifuge in a microfuge at high speed for 10 min.
10. Remove the supernatent by aspiration. Wash the pellet with 70% (v/v) ethanol.
11. Remove the supernatent by aspiration. Invert the tubes on a Kimwipe for 15 min at room temperature to dry off any residual alcohol. (Alternatively, the tubes can be placed in a vacuum dessicator briefly.)
12. Dissolve the pellet in 100 µl TE. (If the DNA solution is turbid, centrifuge for 2 min at 8,000 rpm in a microfuge and use the supernatent. The material in the pellet does not interfere with further manipulation of the DNA so it does not need to be discarded.)

NOTE: This procedure can be scaled up to larger volumes if the centrifugation steps are done in a large centrifuge.

Reagents

Lysis solution (50 mM glucose, 25 mM TrisHCl pH 8, 10 mM EDTA)
9.0 ml	dH$_2$O
0.25 ml	1 M TrisHCl pH 8
0.20 ml	0.5 M Na$_2$EDTA
0.50 ml	20% Glucose

SDS/NaOH (Prepare fresh)
880 µl	dH$_2$O
100 µl	10% SDS
20 µl	10 N NaOH

7.5 M ammonium acetate
Dissolve 57 g ammonium acetate in about 60 ml dH$_2$O.
Bring to 100 ml with dH$_2$0.
Sterilize by passing through a 0.2 µm membrane filter.
Store at room temperature.

7B. Large Scale Plasmid Isolation

1. Grow plasmid containing strain in 2 ml LB.
2. Transfer into 100 ml plasmid broth + 20 µg/ml cam. Grow overnight in a 37°C water bath with vigorous shaking.
3. Spin down cells at 5,000 rpm for 10 min in a GSA rotor.
4. Pour off the supernatant and resuspend the pellet in 20 ml 0.85% NaCl in a SS34 tube. Spin down the cells at 10,000 rpm for 5 min in an SS34 rotor.
5. Resuspend the cell pellet in 8 ml of lysis solution. Mix by inverting the tube. Leave at room temperature for 5 min.
6. Add 16 ml SDS/NaOH. Mix by gently inverting. Leave on ice 5-10 min.
7. Add 12 ml of 7.5 M ammonium acetate. Mix by gently inverting. Leave on ice about 10 min.
8. Spin down the precipitated material at 10,000 rpm for 30 min in an SS34 rotor at 4°C. (The chromosomal DNA and cell debris will pellet).
9. Pour the supernatant through a Kimwipe into a clean SS34 tube. Precipitate the DNA by adding 0.6 volume of room temperature isopropanol. Mix by inverting.
10. Pellet the precipitated DNA at 12,000 rpm in a SS34 rotor for 30 min at room temperature.
11. Pour off the supernatant. Invert the tubes over Kimwipes to drain off the residual alcohol. Resuspend the DNA in 8 ml TE.
12. Add 8.8 g solid CsCl. Mix gently until the CsCl is dissolved.
13. Add 0.8 ml of 10 mg/ml ethidium bromide. Keep covered or turn off the room lights to reduce exposure of the DNA to light. (WEAR GLOVES!)
14. Load into a small quick seal tube (holds 13.5 ml). Add 50% CsCl to completely fill the tube then seal the tube.
15. Place in a Ti70 rotor. Balance with another sample or a quick seal tube filled with 50% CsCl. Cover with spacers. Spin in the ultracentrifuge at 45,000 rpm for at least 20 hrs at 20-25°C. (Depending on the concentration of DNA, up to 40 hrs may be required.)
16. Stick a needle into the top of the tube. Illuminate the tube with a long wave UV lamp. Remove the plasmid band (lower of the two bands) with a syringe with an 18 gauge needle (see diagram below).

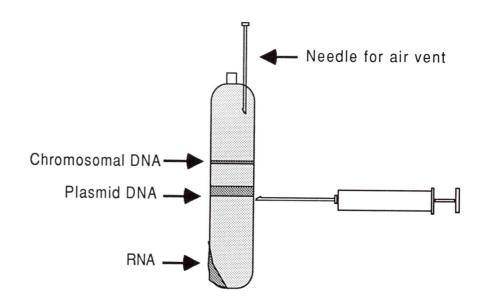

17. Remove the needle, then gently squirt the DNA into a screw-capped glass test tube. (If very pure DNA is required a second centrifugation may be required. If so, add the plasmid band directly to another quick seal tube, fill the tube with 50% CsCl/ethidium bromide, and centrifuge as in step 15.)

18. Extract the ethidium bromide with an equal volume of TE saturated butanol. Mix by inverting. Spin briefly in a clinical centrifuge to separate the layers. Aspirate off the upper layer (butanol and ethidium bromide). Reextract the lower layer until all the color is gone then one more time. At least 5-6 extractions are required.

19. Dialyze against 1000 ml of TE buffer. Change the buffer three times with at least 3 hr between buffer changes. (This will dilute the CsCl to about 10^{-6} of the original concentration).

20. Pour the DNA into a SS34 tube. Add 1/2 volume of 7.5 M ammonium acetate and 2.5x the total volume of 95% ethanol. Place at 4°C overnight.

21. Centrifuge at 10,000 rpm at 4°C for 15 min.

22. Pour off the ethanol and drain over a Kimwipe. Dry briefly in a vacuum dessicator. Resuspend the pelleted DNA in 0.5 ml TE.

7C. Determining the Purity and Concentration of a DNA Solution

The concentration of a DNA solution can be determined by measuring the absorbance of ultraviolet light. A 50 µg/ml solution of double stranded DNA has an OD_{260} = 1. (The value varies slightly with the ratio of G:C to A:T but the error is usually small enough to ignore.) The ratio of absorbance to DNA concentration is linear to an OD = 2, so it is possible to determine the concentration of a pure DNA solution from the OD_{260}.

The purity of a DNA solution can also be determined by measuring the absorbance of ultraviolet light. Due to the characteristic absorption spectra of DNA, the ratio of absorbance at 260 nm/280 nm for a pure solution of double stranded DNA should be between 1.7 and 1.9. Higher ratios are often due to RNA contamination and lower ratios are due to protein or phenol contamination. If the DNA solution is not pure, the OD_{260} will not reflect the actual DNA concentration.

Determine the DNA concentration and purity of a DNA solution by reading the absorbance at 260 and 280 nm. Use quartz cuvettes.

1. Turn on the spectrophotometer and allow it to warm up.
2. Zero the spectrophotometer with 1x TE.
3. Dilute 5 µl DNA into 1 ml TE. Read the OD_{260} and OD_{280}.
4. Calculate the OD_{260}/OD_{280} ratio. If the ratio is between 1.7 and 1.9, calculate the concentration of DNA as follows:

$$\mu g/ml\ DNA = \frac{(50\ \mu g/ml)}{OD_{260}} \times (OD_{260}\ measured) \times (dilution\ factor)$$

7D. Restriction Mapping

Remember the following guidelines for using restriction enzymes:

- Most restriction enzymes lose activity if allowed to warm up (even on ice). Therefore, prepare all reagents except the restriction enzyme. Then remove the enzyme from the freezer and immediately place on ice. Add the enzyme then return it to the freezer as soon as possible.
- Wear gloves and always use a new, sterile pipet tip every time you use a restriction enzyme.
- The volume of restriction enzyme added should be less than 1/10 of the final volume of the reaction mixture.
- Often the amount of enzyme can be reduced if the digestion time is increased. Small aliquots of the sample can be removed during the course of the reaction and run on a minigel to monitor the progress of the digestion. Sometimes a DNA sample will not cut no matter how long you give it because of contaminants in the DNA. Such inhibitory contaminants can usually be removed by another ethanol precipitation or by drop dialysis (Appendix 6A).

A good rule of thumb is to start with relatively inexpensive restriction enzymes that only cut your plasmid 1 or 2 times. Try the following enzymes first: *Bam* HI, *Eco* RI, *Hin* dIII, *Kpn* I, *Pst* I, *Sal* I, *Xba* I, and *Xho* I. Use the 10x restriction enzyme buffers supplied by the manufacturer or the appropriate buffer from Appendix 6A.

1. Digest purified plasmid DNA from potential clones and pBR328 as follows:

dH$_2$O	___ μl
DNA (0.2 μg)	___ μl
10x RE buffer	2 μl
RE	1 μl
Total	20 μl

Also run an uncut plasmid control prepared as above but without adding the restriction enzyme.

2. Mix with a pipetman. Incubate 1-2 hrs at 37°C.
3. Remove 5 μl. Add 5 μl TE and 1 μl Blue II and run on a 0.8% agarose minigel (Appendix 7A). Include Lambda *Hin* dIII standards (1 μl Lambda *Hin* dIII + 9 μl TE + 1 μl Blue II).
4. Stain the gel with ethidium bromide.
5. Photograph the ethidium bromide stained gel as described in Appendix 7D. **CAUTION: WEAR UV PROTECTIVE GLASSES. UV LIGHT FROM THE TRANSILLUMINATOR CAN SERIOUSLY INJURE YOUR EYES WITHOUT EYE PROTECTION.**
6. Determine the size of the restriction fragments from your potential clones. Determine which fragments come from the vector and which fragments come from the cloned DNA by comparison of the pBR328 digest with your clones. For each plasmid, the total size of the fragments in the different digests should be the same. If the total size of one digest is greater than the others, it may be a partial digest. (If any of the smaller bands are brighter than the larger bands you probably have a partial digest). If the total size of the fragments is less in one digest, carefully examine the gel for bright bands that may be doublets. Also, small fragments may run off the gel if they migrate faster than the bromphenol blue tracking dye.

Choose the restriction enzymes that look most useful for your fragment and digest with a second enzyme. If the two enzymes both require the same buffer, digest with both enzymes simultaneously. If the two enzymes require different buffers, the first enzyme should be inactivated before changing the buffer and adding the second enzyme. Some restriction enzymes can be inactivated by heating to 65°C for 20 min (see Appendix 6A). If the first enzyme is inactivated by heating, heat the reaction mixture, adjust the salt concentration of the buffer (Appendix 6B), and add the second enzyme. If the first enzyme is not inactivated by heating, extract the first digest with an equal vol of phenol:chloroform:isoamyl alcohol (25:24:1), ethanol precipitate, and resuspend in the appropriate buffer before doing the second digest (see Appendix 5).

RESULTS
EXPERIMENT 7

1. Indicate the strains used and the predicted gene cloned.
2. Include a photocopy of the agarose gel(s) with a legend describing what was run in each lane.
3. Include a plot of the size of the Lambda *Hind*III standards vs distance migrated.
4. Draw the predicted restriction map of the clone showing the sizes of the restriction fragments.

8

INSERTION MUTAGENESIS OF PLASMID CLONES

The purpose of this experiment is to determine the location of the complementing gene in the plasmid clone.

IDENTIFYING A GENE ON A CLONED FRAGMENT

The simplest way to locate a specific gene in a cloned DNA fragment is to isolate mutations in the cloned gene and restriction map the mutations. Large insertions or deletions in the cloned fragment are usually isolated because they are easiest to detect by restriction mapping. Two approaches are commonly used.

(1) *The in vitro approach.* Using restriction enzymes, deletions may be made in the plasmid clone *in vitro* , then the deleted clones transformed into cells to determine if the gene of interest has been mutated (e.g., by complementation of a chromosomal mutation). If a detailed restriction map of the clone is available, many deletions can be made to narrow down the position of the gene on the cloned fragment. The plasmid clone can be cut with a restriction enzyme that removes a fragment from the clone, then the ends of the plasmid can be religated. Sometimes it is necessary to construct deletions between two different restriction sites: if the two restriction enzymes produce incompatible sticky ends, the sticky ends can be converted to blunt ends prior to ligation (Maniatis et al., 1982).

(2) *The in vivo approach.* Alternatively, transposon insertions can be isolated in the plasmid clones *in vivo* , then the cells can be tested to identify insertions into the gene of interest. The sequence specificity of the transposon used must be nearly random in order to isolate mutations throughout the plasmid clone. The position of the insertions can be identified by restriction mapping. This approach has two advantages. (a) If many insertions are isolated in the cloned gene, it is possible to determine the extent of the gene even if no useful restriction sites are present near the 5' and 3' ends of the gene. (b) By correlating the position of several insertions with the size of the truncated proteins made, it is possible to determine the direction of transcription of the gene.

Tn *1000*

In this experiment, Tn*1000* insertion mutations are isolated in the complementing clones from Experiment 6. Tn*1000* (also called gamma-delta) is a 5.7 kb transposon that is closely related to Tn*3* . It has 35 bp inverted repeats on each end and it encodes transposase and resolvase functions but, unlike Tn*3*, it does not encode antibiotic resistance. Tn*1000* is present on the *E. coli* chromosome and the F-factor, but not on the *S. typhimurium* chromosome (Guyer, 1978).

TRANSPOSITION OF Tn *1000* ONTO PLASMIDS

Transposition of Tn*1000* onto nonconjugative plasmids can be selected by requiring mobilization by the F-factor as shown in Figure 8. The plasmid pBR328 cannot be conjugated from one bacterium to another, even if conjugation functions are provided by an F-factor present in the same cell. However, when both plasmids are present in the same cell, occasionally the Tn*1000* on the F-factor will transpose onto the plasmid. Transposition of Tn*1000* produces an intermediate (called a cointegrate) which have pBR328 integrated into the F-factor DNA with a direct repeat of Tn*1000* at each junction (Figure 8B). If the F-factor carrying a plasmid cointegrate is mated into another cell, the plasmid is passively transferred as well. Recombination between the two copies of Tn*1000* resolves the F-factor and the plasmid which now has an insertion of Tn*1000* (Figure 8C). Thus, by directly selecting for mating of pBR328, every exconjugant will have a Tn*1000* insertion in pBR328. Each pBR328 clone will have a different insertion somewhere in the plasmid, so individual colonies must be screened for the mutant phenotype.

References

de Bruijn, F., and J. Lupski. 1984. The use of transposon Tn5 mutagenesis in the rapid generation of correlated physical and genetic maps of DNA segments cloned into multicopy plasmids -- a review. *Gene 27*: 131-149.

Guyer, M. 1978. The gamma-delta sequence of F is an insertion sequence. *J. Mol. Biol. 126*: 347-365.

Guyer, M. 1983. Uses of the transposon gamma-delta in the analysis of cloned genes. *Methods Enzymol. 101*: 362-369.

Maniatis, T., E. Fritsch, and J. Sambrook. 1982. *Molecular Cloning: A Laboratory Manual*, pp. 398-400. Cold Spring Harbor Laboratory, NY.

Figure 8. Isolation of Tn1000 insertion mutants in plasmid clones

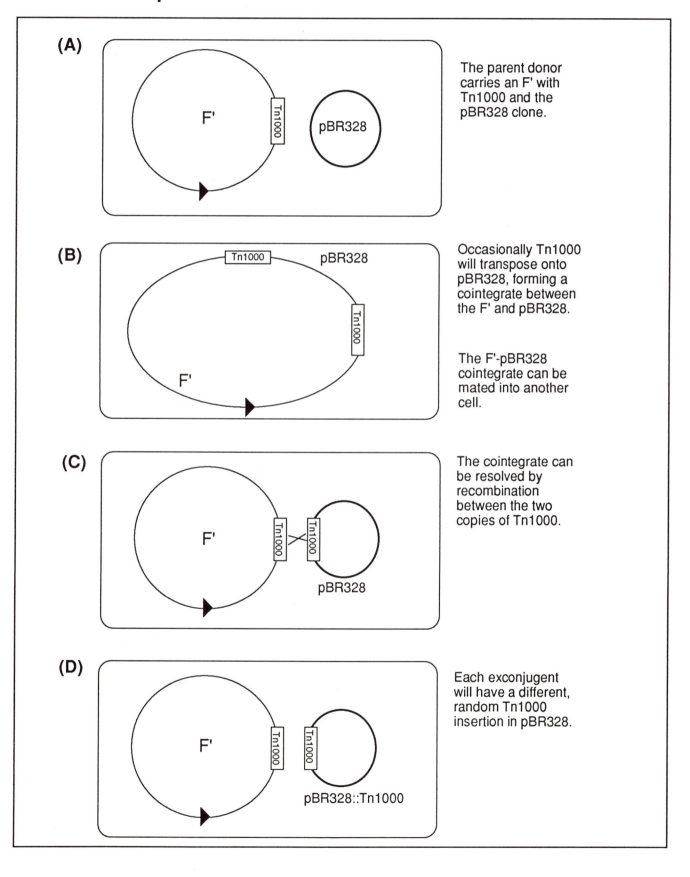

(A) The parent donor carries an F' with Tn1000 and the pBR328 clone.

(B) Occasionally Tn1000 will transpose onto pBR328, forming a cointegrate between the F' and pBR328.

The F'-pBR328 cointegrate can be mated into another cell.

(C) The cointegrate can be resolved by recombination between the two copies of Tn1000.

(D) Each exconjugent will have a different, random Tn1000 insertion in pBR328.

1. Grow phage P22 on the strain with the complementing pBR328 clone as follows:
   a. Grow an overnight of your strain in 1 ml NB in a 37°C shaker.
   b. Add 4 ml P22 broth.
   c. Incubate 8-16 hrs in the 37°C shaker.
   d. Spin down 20 min in a clinical centrifuge to pellet the cell debris.
   e. Pour the supernatant into a screw capped test tube. Add several drops of chloroform and vortex. Store at 4°C.

2. Using the phage from step #1, transduce MS1974 [del(*his*)/F'42 *finP301* Lac$^+$ Tn*10*] selecting for growth on NB + Cam plates.
   a. Grow the recipient strain overnight in 2 ml NB.
   b. Dilute the phage 1/50 in sterile 0.85% NaCl.
   c. Mix the cells and diluted phage NB plates as follows:

plate	ml cells	ml phage	
A	0.1	—	cell control
B	0.1	0.05	
C	0.1	0.10	
D	0.1	0.20	
E	—	0.20	phage control

   d. Spread the plate with an alcohol-flamed glass spreader.
   e. Incubate the plates 3-4 hrs at 37°C to allow phenotypic expression.
   f. Replica plate onto NB + chloramphenicol plates and incubate overnight at 37°C.
   g. Clean up a transductant on green plates and cross-streak against phage H5 (see page 25).

3. Grow this His$^-$/F'Lac$^+$Tn*10* / pBR328(Aux$^+$) donor and your original auxotrophic MudJ mutant (recipient) overnight in 1 ml NB.

4. Mate the donor and recipient as follows:
   a. Mix the donor and recipient cultures in separate sterile microfuge tubes as shown below.

Tube	Donor cells	Recipient cells	
A	50 µl	—	Donor control
B	—	50 µl	Recipient control
C	50 µl	50 µl	

   b. Incubate 1 hr at 37°C.
   c. Spot each mating mixture on an E + glucose + Cam plate spread with 0.1 ml of the required auxotrophic supplement. (Requiring His$^+$ selects against the donor and requiring Cam^r selects against the recipient.) Spot the mating mixture containing donor and recipient in the center of the plate and spot the controls on each side as shown on page 9.
   d. Allow the drops to dry onto the plate, then streak the plate from the drops out and incubate at 37°C for 1-2 days.

4. Patch the colonies on the same medium. Incubate overnight at 37°C.
5. Check for insertions in or near the cloned gene:
   a. Replica plate onto E + glucose + Cam plates and NB + Amp plates. Any colonies with a Tn*1000* insertion in the cloned complementing gene will not grow unless supplemented with the auxotrophic requirement. Any colonies with a Tn*1000* insertion in the Amp gene of pBR328 will not grow on the NB + Amp plates. (Amp is a "hot spot" for Tn*1000* insertions, so this is a good positive control.)
   b. Purify any plasmids with the desired insertions (Experiment 7A) and digest with appropriate restriction enzymes to determine the location of the insertion sequence (Experiment 7D). A restriction map of Tn*1000* is shown below.

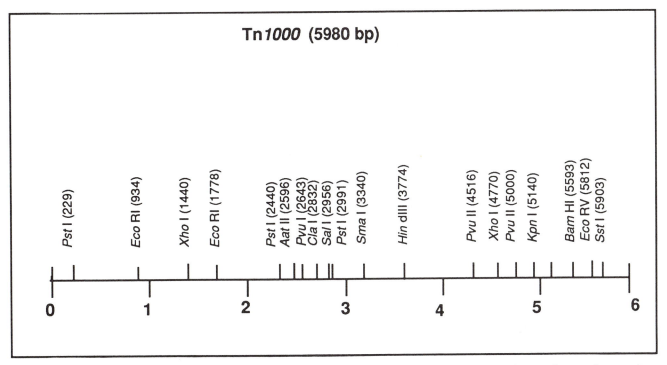

RESULTS
EXPERIMENT 8

1. Include a table showing how many colonies were obtained on each transduction plate.
2. Indicate the results of the mating. (How many colonies were obtained in the donor + recipient spot? Did any colonies grow in the control spots?)
3. Indicate how many Tn*1000* insertion mutations were obtained in the cloned auxotrophic gene and in the Amp gene.
4. Draw a restriction map of the plasmid indicating the position of the Tn*1000* insertions. (If your restriction mapping data is good enough, you should be able to indicate the orientation of the Tn*1000* insertion on the plasmid as well.)

EXPRESSION OF CLONED GENE PRODUCTS

The purpose of this experiment is to analyze the proteins encoded by plasmid genes in maxi-cells.

UV DAMAGE

Maxi-cells result from UV irradiation of *recA* mutants (Sancar et al., 1979; Silhavy et al., 1984). UV damages DNA primarily by forming thymine dimers (T-T) which block DNA replication. Thymine dimers can be repaired by four mechanisms (Freifelder, 1987; Walker, 1984):

(1) *Photoreactivation.* The enzyme photolyase (*phr*) binds T-T dimers. When the enzyme-DNA complex is excited by 300-600 nm light, it breaks the dimers.
(2) *Excision repair.* A repair endonuclease (encoded by the *uvr* genes) recognizes the distortion produced by a T-T dimer and cuts the DNA on both sides of the dimer, leaving a 3'-OH group. DNA polymerase I then synthesizes a new strand from the complementary strand, repairing the strand that had the T-T dimer.
(3) *Recombination repair.* DNA polymerase cannot replicate DNA across a T-T dimer. Under these conditions, somehow DNA polymerase "jumps over" the dimer and begins unprimed DNA synthesis, resulting in large gaps in the new strand directly across from the T-T dimers. These gaps can be repaired by *recA* mediated recombination with a sister strand. The recombinant strands can then serve as templates for DNA replication.
(4) *SOS:-repair* A *recA* dependent repair system mediated by the *umuC* and *umuD* gene products induces error prone replication across from T-T dimers. SOS-repair probably accounts for most UV-induced mutations.

Of these four mechanisms, the two *recA* dependent repair systems are most important for survival after UV irradiation.

MAXI-CELLS

When *E. coli* is given a heavy dose of UV light the chromosome is extensively damaged. In *recA* mutants neither recombinational repair nor SOS-repair can occur. In addition, if the cells are kept in dim light, photoreactivation cannot occur. If the gaps are not repaired, the chromosomal DNA is degraded by nucleases. Under these conditions cell division is inhibited but the cells continue to elongate, forming long filamentous cells or maxi-cells (see Figure 9-1).

DNA on multicopy plasmids is much more likely to survive UV irradiation for two reasons: (1) the target size of a plasmid is much smaller than the chromosome, so the plasmid is less likely to be "hit" by UV; and (2) multiple copies of the plasmid are present (usually 10-50 copies per cell) so it is unlikely every copy of the plasmid will be hit. Thus, with an optimal dose of UV the chromosome is degraded but plasmids remain intact. Any cells that survive UV irradiation (i.e., if the chromosome remains

Figure 9-1. Maxicells

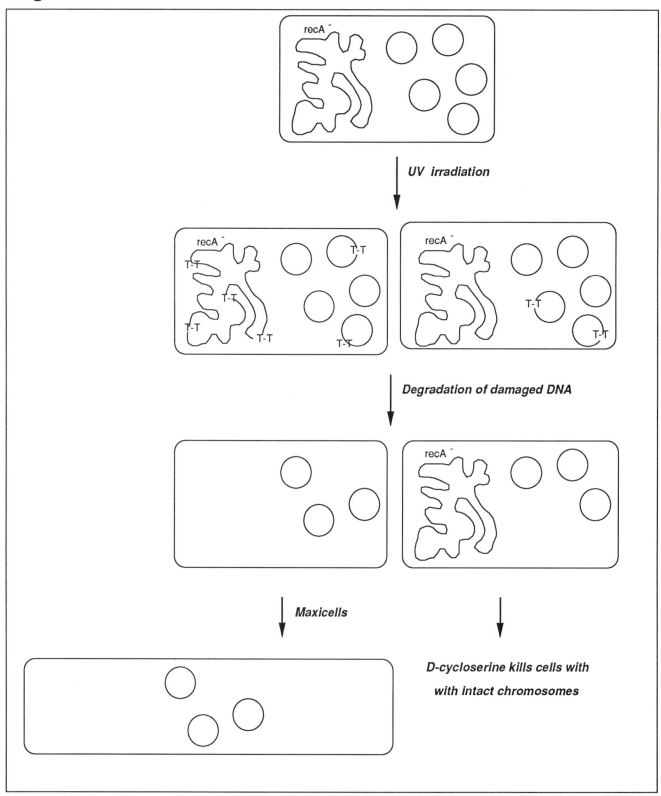

intact) are killed with the antibiotic cycloserine. Cycloserine kills growing cells by preventing racemization of L-alanine to D-alanine, which blocks cross-linking of the cell wall peptidoglycan. After this treatment ^{35}S-methionine is added to label any newly synthesized proteins. Since the chromosomal DNA is degraded and the half-life of bacterial mRNAs is relatively short, only the plasmid encoded genes are transcribed and translated into new proteins. The total cell proteins are separated on an SDS-polyacrylamide gel and the ^{35}S-labeled proteins visualized by autoradiography. By comparing the proteins produced by plasmid clones with the parent plasmid it is possible to determine the size of the polypeptides encoded on the cloned DNA.

In order to study the plasmid proteins expressed in maxi-cells, the plasmids will be transformed into an *E. coli* strain that is *recA* (to allow formation of maxi-cells), *metB* (to increase the incorporation of exogenous ^{35}S-methionine), and *hsdR* (restriction⁻ modification⁺ to allow transformation with DNA from *S. typhimurium*).

Hsd MUTANTS

Restriction enzymes are part of restriction-modification systems which may play an important role in protecting bacteria from invasion by foreign DNA. A modification enzyme methylates specific sequences in the host DNA and a restriction endonuclease cleaves DNA if the appropriate site is not methylated. Although the restriction-modification system is useful for the bacterium, it makes life tougher for a molecular biologist trying to clone foreign DNA. *E. coli* makes the Type I restriction enzyme *Eco* K. This enzyme has two subunits. One subunit (encoded by *hsdS*) recognizes the sequence AACNNNNNNNNTGCT. Another subunit (encoded by *hsd* R) cuts the DNA at a variable distance from this site. In order to avoid digesting its own DNA, the *Eco* K recognition site is modified in *E. coli* K-12. When the adenines in this site are methylated by the enzyme *Eco* K methylase (encoded by *hsdM*), it is no longer recognized by *Eco* K .

When foreign DNA is transformed into an Hsd⁺ *E. coli* strain there is a "race" between the endonuclease and the methylase to any *Eco* K sites. The endonuclease wins over 90% of the time. Thus, even if a plasmid only has one *Eco* K site, less than 1/10 as many transformants will be obtained in a *hsdR* ⁺ host compared to *hsdR* ⁻. Therefore, a *hsdR* ⁻ strain is used for cloning *S. typhimurium* DNA into *E. coli* .

TRANSFORMATION

E. coli and *S. typhimurium* are not efficiently transformed with exogenous DNA unless treated to become competent for transformation. Two protocols for preparing competent cells are described. The most common method of preparing competent cells is by hypotonic Ca⁺⁺ shock (Experiment 7A). To prepare competent cells by this method, an early log phase culture is centrifuged and resuspended in a cold hypotonic CaCl₂ solution. When DNA is added to these cells it forms a calcium-DNA complex that adsorbs to the cell surface. The cells are then briefly warmed (heat shocked), which allows transport of the DNA into the cell. If selecting for antibiotic resistance, the cells are often grown briefly in a nonselective medium to allow expression of the antibiotic resistance gene prior to plating on selective medium. A newer procedure that is quicker and easier but works just as well is described in Experiment 7B (Chung and Miller, 1988). In the second technique a polyethylene glycol (PEG)-DNA complex adsorbs to the cell surface. The DMSO apparently facilitates entry of the DNA into the cell because heat shock is not required. Both techniques yield transformation efficiencies of about 2×10^8 transformants per µg DNA.

Several factors affect the tranformation frequency. (1) Rapidly growing early log phase cells must be used. Using a culture that is not well aerated or cells grown to high density decreases the transformation efficiency. (2) Competent cells are very fragile so it is important to keep the cells cold and avoid vigorous mixing. Even briefly allowing the cells to warm up during the treatment decreases the transformation efficiency. (3) For unknown reasons, different strains have different inherent transformation efficiencies. The *E. coli* strain used for these experiments transforms very well. (4) Competent cells can be stored frozen in DMSO at -70°C. In our experience freezing decreases the transformation frequency of competent cells prepared by the CaCl$_2$ procotol at least 10-fold, but does not significantly affect the transformation efficiency of competent cells prepared by the PEG-DMSO protocol.

SDS-POLYACRYLAMIDE GELS

Sodium dodecyl sulfate or SDS [CH$_3$–(CH$_2$)$_{11}$-SO$_4$ Na$^+$] is an anionic detergent. When proteins are boiled in the presence of SDS, the hydrophobic portion of SDS (dodecyl sulfate) interacts with the protein and unfolds it into a long rod shape (see below).

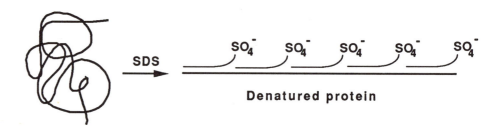

If a sulfhydryl reagent like dithiothreitol (DTT) or ß-mercaptoethanol is included to reduce disulfide bonds, most proteins are fully denatured by SDS. Most proteins bind one molecule of dodecyl sulfate per amino acid (Scopes, 1982). Each molecule of dodecyl sulfate has a negative charge, so for most proteins saturated with SDS the net negative charge of the SDS "masks" the intrinsic charge of the protein. Hence, when saturated with SDS, the net negative charge per molecular weight is identical for most proteins. Since most proteins denatured with SDS have the same shape and the same charge/molecular weight, separation of proteins on SDS-polyacrylamide gels is solely due to the molecular sieving through the pores of the polyacrylamide gel and the migration of the SDS-protein complexes is directly proportional to the log$_{10}$ of the molecular weight of the protein. [Very hydrophobic membrane proteins are notable exceptions. Often very hydrophobic proteins bind more SDS, resulting in a greater negative charge per molecular weight. This causes the protein to migrate faster during electrophoresis, resulting in a lower apparent molecular weight (see Hahn et al., 1988)]. Thus, by comparing the mobility of an unknown protein with a set of standard proteins of known molecular weight, the molecular weight of the unknown protein can usually be determined quite accurately.

ELECTROPHORESIS

Use of a discontinuous electrophoresis system improves the resolution of proteins separated on SDS-polyacrylamide gels. Discontinuous electrophoresis involves running a gel composed of two different layers: the proteins initially pass through a low concentration acrylamide "stacking gel " that is layered on top of a "running gel." Because of the difference between the ion concentrations in the stacking gel and the electrophoresis buffer, proteins concentrate into a sharp band in the stacking gel (see Ausubel et al., 1988). Thus, the proteins enter the running gel as a narrow, concentrated band. The running gel has a higher acrylamide concentration which acts as a molecular sieve, separating proteins of different molecular weights.

References

Ames, G. F. 1974. Resolution of bacterial proteins by polyacrylamide gel electrophoresis on slabs. *J. Biol. Chem. 249*: 634-644.

Ausubel, F., R. Brent, R. Kingston, D. Moore, J. Smith, J. Seidman, and K. Struhl. 1987. *Current Protocols in Molecular Biology*, p. 1027. John Wiley and Sons, NY.

Calhoun, D., and J. Gray. 1981. Detection of proteins encoded by cloned DNA segments. *BRL Focus*, v. 3.

Chung, C. and R. Miller. 1988. A rapid and convient method for the preparation and storage of competent bacterial cells. *Nucleic Acids Res. 16*:632.

Davis, R., D. Botstein, and J. Roth. 1980. *Advanced Bacterial Genetics*, pp. 140-141. Cold Spring Harbor Laboratory, NY.

Freifelder, D. 1987. *Molecular Biology*, Second Edition, pp. 282-292. Jones and Bartlett Publishers, Boston, MA.

Hahn, D., R. Myers, C. Kent, and S. Maloy. 1988. Regulation of proline utilization in *Salmonella typhimurium*: molecular characterization of the *put* operon and DNA sequence of the *put* control region. *Mol. Gen. Genet. 213*: 124-133.

Sancar, A., A. Hack, and W. Rupp. 1979. Simple method for identification of plasmid coded proteins. *J. Bacteriol. 137*: 692-693.

Scopes, R. 1982. *Protein Purification*, pp. 248-249. Springer-Verlag, NY.

Silhavy, T., M. Berman, and L. Enquist. 1984. *Experiments with Gene Fusions*, pp. 213-214. Cold Spring Harbor Laboratory, NY.

Walker, G. 1984. Mutagenesis and inducible responses to deoxyribonucleic acid damage in *Escherichia coli*. *Microbiol. Rev. 48*: 60-93.

9A. Transformation of Plasmid DNA (CaCl$_2$ procedure)

1. Grow EM158 [*E. coli* del(*recA-srl*) *srl*::Tn*10 zfi*::Tn*10* dCam *hsdR metB*] overnight in 2 ml NB at 37°C.
2. Subculture 0.2 ml into 20 ml NB in a Klett flask. Grow to early log phase (about 50 Klett units) on a 37°C shaker.
3. Cool cells on ice 20-30 min.
4. Spin down cells 5 min at 5000 rpm in SS34 rotor at 4°C.
5. Gently resuspend pellet in 10 ml ice cold 50 mM CaCl$_2$. DO NOT VORTEX. Keep on ice for about 30 min.
6. Spin down cells 5 min at 5000 rpm in SS34 rotor at 4°C.
7. Gently resuspend cells in 1.0 ml ice cold 50 mM CaCl$_2$. DO NOT VORTEX. Keep on ice at least 30 min. The cells are competent for transformation after this step.
8. Separate transformations should be done with your original clone and any Tn*1000* insertion mutants you isolated. In glass test tubes on ice add aliquots of purified plasmid DNA and competent cells as shown below:

Tube	Plasmid DNA	Competent cells	
A	—	0.1 ml	Cell control
B	1 μl	0.1 ml	
C	10 μl	0.1 ml	
D	10 μl	—	DNA control

9. Keep on ice for 30 min.
10. Heat pulse by placing in a 37°C water bath for 2 min then return to ice for 30 min.
11. Add 0.2 ml NB and incubate in a 37°C shaker for 30 min.
12. Spread directly on NB + Amp plates. Incubate the plates upside-down at 37°C overnight.
13. Restreak the transformants on NB + Amp plates. Incubate the plates upside down at 37°C overnight then store at 4°C.

9B. Transformation of Plasmid DNA (PEG / DMSO protocol)

1. Grow EM158 [*E. coli* del(*recA-srl*) *srl*::Tn*10* *zfi*::Tn*10* dCam *hsdR* *metB*l overnight in 2 ml NB at 37°C.
2. Subculture 0.2 ml into 20 ml NB in a Klett flask. Grow to early log phase (about 50 Klett units) on a 37°C shaker.
3. Centrifuge in the SS34 rotor at 5000 rpm for 5 min at 4°C .
4. Resuspend the cell pellet in 1/10 volume cold TSB.
5. Incubate on ice for 10 min. (After this step the cells are competent for transformation. Competent cells prepared by this method can be frozen in a dry ice-ethanol bath and stored at -70°C, then thawed on ice without significant loss of transformation efficiency.)
6. Separate transformations should be done with your original clone and any Tn*1000* insertion mutants you isolated. In tubes on ice add aliquots of purified plasmid DNA and competent cells as shown below:

Tube	Plasmid DNA[a]	Competent cells	
A	—	0.1 ml	Cell control
B	1 µl	0.1 ml	
C	10 µl	0.1 ml	
D	10 µl	—	DNA control

[a] About 100 pg of plasmid DNA is ideal.

7. Leave on ice for 5-10 min.
8. Add 0.9 ml NB to each tube and grow for 60 min in the 37°C shaker to allow phenotypic expression.
9. Spread on NB + Amp plates. Incubate overnight at 37°C.

Reagent

TSB
100	ml	NB
10	g	PEG (MW = 3,350)
5	ml	DMSO
0.1	ml	1 M MgSO$_4$
0.1	ml	MgCl$_2$

9C. Labeling Plasmids Encoded in Maxicells

1. Start a single colony of the plasmid carrying strains and the isogenic control without plasmids (EM158) in 2 ml E medium + 0.2% DAA + 0.4% succinate + 0.1% thiamine. Grow overnight on a 30°C shaker.
2. Subculture 0.1 ml of the cells into 5 ml of the same medium in a Klett flask and grow to 100-120 Klett units on the 37°C shaker.
3. Spin down the cells 5 min at 10,000 rpm in the SS34 rotor.
4. Resuspend the pellet in 20 ml of 10 mM $MgSO_4$.

NOTE: CARRY OUT STEPS 5-9 IN LOW LIGHT TO PREVENT PHOTO-REAC-TIVATION.

5. Pour the cells into a sterile plastic petri dish. Remove the cover and irradiate the cells with UV at 200 μW / cm^2 for 30 sec.
6. Spin down the irradiated cells at 10,000 rpm for 5 min in the SS34 rotor.
7. Resuspend the pellet in 10 ml E + DAA + succinate + thiamine.
8. Spread 0.1 ml on an LB plate. Incubate the plate at 37°C overnight. (This is a control for the efficiency of UV killing. There should be less than 1% survival.)
9. Transfer 5 ml of the cell suspension to a foil-wrapped test tube.
10. Incubate the foil covered tube in a 37°C shaker for 1 hr to allow recovery of any remaining viable cells.
11. Add 25 μl of 50 mg/ml cycloserine to the tube. Place the tube in a 37°C shaker overnight.
12. Spread 0.1 ml on an LB plate. Incubate the plate at 37°C overnight. (This is a control for the efficiency of cycloserine killing. There should be few or no survivors).
13. Spin down the cells at 10,000 rpm for 5 min in the SS34 rotor. Resuspend the cell pellet in 5 ml of 10 mM $MgSO_4$. Vortex and recentrifuge.
14. Resuspend the cell pellet in 1 ml of E + DAA + succinate + thiamine medium in a microfuge tube. Incubate at 37°C for 1 hr.

The rest of this experiment involves radioactivity. WEAR GLOVES AND A LAB COAT.

*15. Add 10 μl [^{35}S]methionine (to 25-100 μCi/ml).
*16. Incubate 30 min. Most of the label is incorporated in the first few minutes.
*17. Spin for 5 min in the microfuge. Remove the supernatant and discard in the radioactive waste.
*18. Resuspend the cell pellet in 0.5 ml of 10 mM $MgSO_4$.
*19. Spin for 5 min in the microfuge.
*20. Pour off the supernatant into the radioactive waste.
*21. Wash the pellet with 0.5 ml 10 mM $MgSO_4$ as described in steps 19-21.
*22. Resuspend the final cell pellet in 0.1 ml of SDS sample buffer and store at -20°C until use.

Reagents

Defined amino acids minus methionine (DAA)
Mix 5 ml of each of the following amino acid stock solutions:

alanine	glycine	leucine	threonine
arginine	glutamate	lysine	tryptophan
aspartate	histidine	proline	tyrosine
cysteine	isoleucine	serine	valine

Concentrations of the amino acid stock solutions are shown in Appendix 4. Add 40 ml DAA to 210 ml E medium to obtain the final concentration of amino acids shown in Appendix 4.

D-cycloserine (Prepare fresh)
50 mg D-cycloserine
1 ml sterile dH_2O

SDS-Sample Buffer (4% SDS, 62.5 mM Tris pH 6.8, 10% Glycerol, 8 M Urea, 0.010% Bromphenol Blue)

1.0 g	SDS
2.5 ml	0.625 M Tris pH 6.8
2.0 ml	Glycerol
2.0 mg	Bromphenol Blue
12.0 g	Urea
0.5 g	DTT

Bring to 25 ml with ddH_2O.
Aliquot into microfuge tubes and store at -20°C.

9D. SDS-Polyacrylamide Gel Electrophoresis of Proteins

CAUTION - Unpolymerized acrylamide is a neurotoxin. Wear gloves and do NOT mouth-pipet.

1. Assemble gel plates with spacers and clamps (see Figure 9-2). Seal the outside edges with melted 1% agarose.
2. Prepare the SDS-polyacrylamide running gel as described below.
    a. In a small vacuum flask, mix the appropriate volume of acrylamide /bis-acrylamide and buffer according to the following table.

12% Acrylamide Running Gel

	Large gel	Mini-gel
30% Acrylamide / 0.8% bis	9.4 ml	4.2 ml
Running gel buffer	13.0 ml	5.8 ml
Ammonium persulfate	150 µl	67 µl
TEMED	7 µl	3 µl

Large gel = 16 cm x 20 cm x 0.8 mm
Mini-gel = 8 cm x 10 cm x 1.5 mm

b. Warm the solution to room temperature. Degas by placing a rubber stopper over the top of the vacuum flask, connect the flask to a vacuum, and gently swirl until the solution quits bubbling (see Appendix 7B).
c. Add the ammonium persulfate and TEMED. Swirl and immediately pour the running gel as shown in Figure 9-2.
d. Using a Pasteur pipet, gently overlay the top of the gel with a small amount of water saturated isobutanol.
e. Allow to polymerize at least 1 hr. Leave a small amount of acrylamide solution in a Pasteur pipet to check the polymerization.
3. Prepare the SDS-polyacrylamide stacking gel as described below:
    a. In a small vacuum flask, mix the appropriate volume of acrylamide/bis and buffer according to the table below.

6% Acrylamide Stacking Gel

	Large gel	Mini-gel
30% Acrylamide/0.8% bis	2 ml	1 ml
Stacking gel buffer	8 ml	4 ml
Ammonium persulfate	120 µl	60 µl
TEMED	7 µl	4 µl

Figure 9-2. SDS-polyacrylamide gels

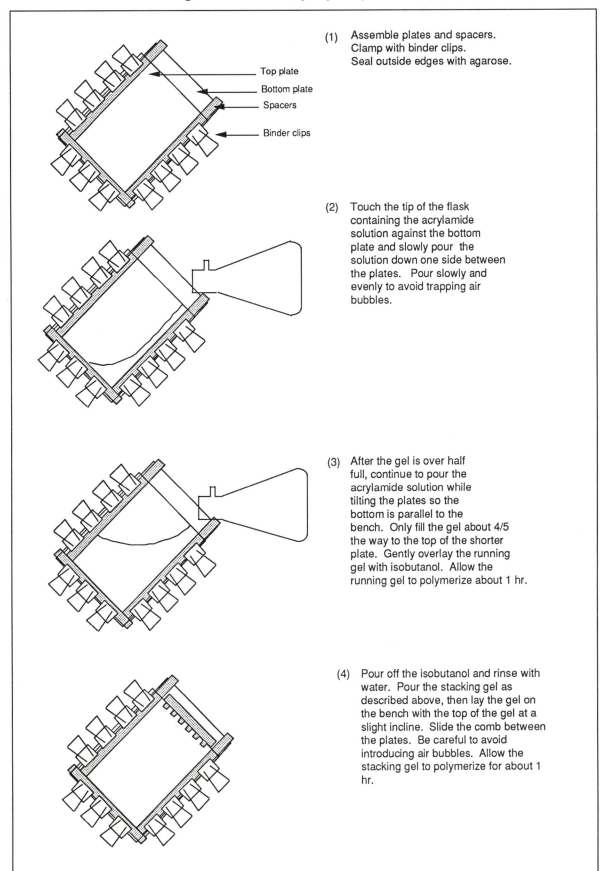

(1)   Assemble plates and spacers. Clamp with binder clips. Seal outside edges with agarose.

Top plate
Bottom plate
Spacers
Binder clips

(2)   Touch the tip of the flask containing the acrylamide solution against the bottom plate and slowly pour the solution down one side between the plates. Pour slowly and evenly to avoid trapping air bubbles.

(3)   After the gel is over half full, continue to pour the acrylamide solution while tilting the plates so the bottom is parallel to the bench. Only fill the gel about 4/5 the way to the top of the shorter plate. Gently overlay the running gel with isobutanol. Allow the running gel to polymerize about 1 hr.

(4)   Pour off the isobutanol and rinse with water. Pour the stacking gel as described above, then lay the gel on the bench with the top of the gel at a slight incline. Slide the comb between the plates. Be careful to avoid introducing air bubbles. Allow the stacking gel to polymerize for about 1 hr.

    b.   Pour the overlay off the running gel. Rinse the top of the gel with a small amount of ddH$_2$O to remove any residual isobutanol.

    c.   Degas the stacking gel solution as described for the running gel.

    d.   Add the ammonium persulfate and TEMED. Swirl and immediately pour on top of the running gel as shown in Figure 9-2.

    e.   Insert the comb. Be careful not to introduce air bubbles.

    f.   Allow to polymerize about 30 min. Leave a small amount of poly-acrylamide solution in a Pasteur pipet to check the polymerization.

4.   Remove the comb. Clamp the gel to the electrophoresis apparatus with neoprene spacers between the plates and the upper buffer reservoir (see Figure 9-3). Seal the edge between the inside plate and the electrophoresis chamber with 1% agarose.

5.   Fill both the electrophoresis reservoirs with electrode buffer. Using a Pasteur pipet with a rubber bulb, flush out the wells with electrode buffer from the upper reservoir. Using a Pasteur pipet bent into a small V at the tip, flush any air out of the slot left by the bottom spacer with buffer from the bottom reservoir.

Figure 9-3. Acrylamide gel electrophoresis set-up

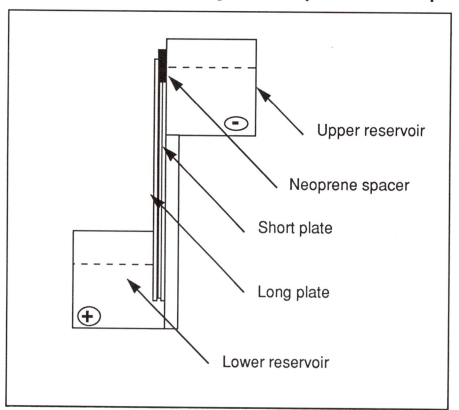

Upper reservoir

Neoprene spacer

Short plate

Long plate

Lower reservoir

CAUTION: The remaining steps involve radioactivity. WEAR GLOVES.

*6. Remove the ^{35}S-labeled proteins from the freezer. Place in a 95°C heating block for 5-10 min. (This thaws the samples and denatures the proteins).

*7. Spin for 2 min in a microfuge to pellet any DNA or cell wall debris.

*8. Gently load 40 µl of each sample into the bottom of the wells with a Hamilton syringe. Load the molecular weight standards in one lane also.

*9. Run the gel at 40 mA constant current. Turn off the power when the dye front reaches bottom of the gel — about 1.5 hrs for mini-gels or about 4-5 hrs for large gels. (Large gels can be run for 16 hrs at 10 mA if it is more convenient.)

*10. Separate the glass plates by gently prying them apart with a spatula. Stain the gel by soaking in the Commassie blue stain solution for about 1 hr.

*11. Place one hand on top of the gel and gently pour off the stain solution into the radioactive waste. Add the destain solution. Destain for 1 hr to overnight.

*12. Pour the destain solution into the radioactive waste, leaving the gel on a glass plate. Briefly rinse the gel in dH$_2$O. Slowly lay a sheet of filter paper over the gel avoiding wrinkles. Slowly pull the paper off the plate — the gel should stick to the filter paper. Cover the gel with plastic wrap.

*13. Dry on a gel dryer at 80°C for about 1 hr. (The time required to dry the gel will depend on the gel dryer and vacuum used.) Leave the vacuum on about 15 min longer than the heater to keep the gels from curling (make sure there is a cold trap between the gel dryer and the vacuum pump).

Reagents

Acrylamide:Bis-acrylamide Stock (30:0.8)
      30.0 g   Acrylamide
       0.8 g   Bis-acrylamide
      Dissolve in 80 ml ddH$_2$0.
      Bring to 100 ml with ddH$_2$0.
      Filter through a 0.2 µm membrane filter.
      Store in a brown bottle at 4°C.

Running Gel Buffer (67 mM Tris, 0.17% SDS, 0.34% NaCl, 3.45% glucose — Addition of NaCl and glucose to the running gel buffer improves the resolution of low molecular weight proteins.)
      79 ml   2 M Tris pH 8.8
       4 ml   10% SDS
      92 ml   0.85% NaCl
      40 ml   20% Glucose
      20 ml   ddH$_2$O

Ammonium Persulfate (100 mg/ml)
      0.5 g Ammonium persulfate
      Dissolve in 5 ml ddH$_2$O.
      Place 200 µl aliquots in microfuge tubes and store at -20°C.

Isobutanol (Water saturated)
      Pour some isobutanol into a brown bottle. Add about 10 ml ddH$_2$O.
      Mix thoroughly and store at room temperature.

Stacking Gel Buffer (78 mM Tris pH 6.8, 1.25% SDS)
    20   ml   0.625 M Tris pH 6.8
    20   ml   10% SDS
  120   ml   ddH$_2$O

Electrode Buffer
   12.0 g   Tris base
    4.0 g   Sodium dodecyl sulfate
  57.7 g   Ultra-pure glycine
4000 ml   ddH$_2$O
Adjust to pH 8.3 with HCl.

SDS-Sample Buffer (4% SDS, 62.5 mM Tris pH 6.8, 10% Glycerol, 8 M Urea, 0.010% Bromphenol Blue)
    1.0 g   SDS
   2.5 ml   0.625 M Tris pH 6.8
   2.5 ml   Glycerol
   2.0 mg   Bromphenol Blue
  12.0 g   Urea
   0.5 g   DTT
Bring to 25 ml with ddH$_2$O.
Aliquot into microfuge tubes and store at -20°C.

SDS-PAGE Molecular Weight Standards
    Add sample buffer to the molecular weight standards to give the final concentration recommended by the supplier. Store the standards at -20°C until use.

Coomassie Blue Stain (20% Methanol, 10% Acetic acid, 0.1% Coomassie blue)
   500   ml   Methanol
   250   ml   Glacial acetic acid
 1750   ml   ddH$_2$O
   2.5 g   Coomassie Blue R-250
Store at room temperature. The stain solutions can be reused 5-6 times.

Destain (10% Methanol, 10% Acetic acid)
   250 ml   Methanol
   250 ml   Glacial acetic acid
  2000 ml   ddH$_2$O

9E. Autoradiography

*1.   Remove the Saran wrap from the dried gel. Place the gel in an X-ray film holder. Remove a sheet of Kodak X-Omat AR film in the darkroom with the lights out and the safelight on. Replace the cover on the film box. Lay one sheet of film on top of the dried gel. Close the film holder and make sure it is "light tight" before turning the lights on. Mark the outside of the film holder with a radiation label, indicating the isotope (^{35}S), the date, and your name.

2.   Leave the film holder at room temperature. The time required to expose the film depends upon the amount of ^{35}S that was incorporated and the translation efficiency of the cloned gene. Usually 48-72 hrs works well.

3.   Develop the film in the darkroom as follows:
   a.   Remove the film in the darkroom with the lights out and the safelight on.
   b.   Submerge in Kodak X-ray developer and agitate intermittently for 2 min.
   c.   Rinse in water for about 30 sec.
   d.   Fix for 5 min in Kodak rapid fixer.
   e.   Rinse in running water for 15 min.
   f.   Hang the film by one edge to allow it to air dry.

   The blot can be reexposed to film if necessary. Adjust the reexposure time based on the initial intensity of the autoradiogram.

4.   Determine the approximate molecular weight of the radiolabeled proteins as follows:
   a.   Measure the distance the standards migrated on the stained gel. Plot the log of the molecular weight vs distance migrated. (A sample standard curve is shown in Figure 9-4).
   b.   Measure the distance the radioactive proteins migrated on the autoradiogram and determine the size of these proteins by comparison with the standard curve. The antibiotic resistance proteins expressed by pBR328 are useful controls for comparing the autoradiogram with the standard curve:
      Amp (ß-lactamase) = 28,000 + 31,000 daltons
      Cam (chloramphenicol transacetylase) = 23,000 daltons
      Tet = 37,000 daltons.

**RESULTS
EXPERIMENT 9**

1.   Describe the plasmids used.
2.   Indicate the number of transformants obtained.
3.   Indicate the number of colonies on the "UV killing control" plate.
4.   Indicate the number of colonies on the "cycloserine control" plate.
5.   Include the autoradiogram of the SDS-polyacrylamide gel.
6.   Include a plot of the molecular weight of the standards vs distance migrated on the Coomassie stained gel.
7.   Calculate the sizes of proteins expressed from the plasmid clones.

Figure 9-4. SDS-Polyacrylamide gel protein standards

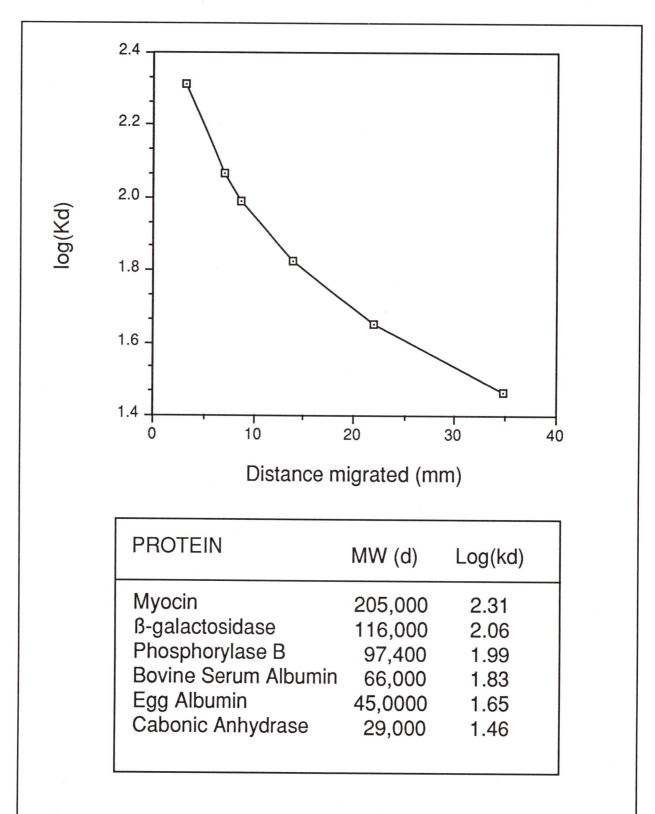

PROTEIN	MW (d)	Log(kd)
Myocin	205,000	2.31
ß-galactosidase	116,000	2.06
Phosphorylase B	97,400	1.99
Bovine Serum Albumin	66,000	1.83
Egg Albumin	45,0000	1.65
Cabonic Anhydrase	29,000	1.46

10

SUBCLONING DNA FRAGMENTS ONTO M13 VECTORS

The purpose of this experiment is to subclone DNA fragments onto M13 vectors to obtain single stranded DNA for sequence analysis.

BIOLOGY OF M13 PHAGE

M13 is a single stranded filamentous phage. M13 adsorbs to the F-pilus (hence it only infects cells that contain an F–factor) and injects its single stranded circular genome into the cell. Once inside the cell, the single stranded phage DNA is replicated into the double stranded replicative form (RF). The RF undergoes Oreplication until about 200 copies are produced per cell. Then one strand of the RF (the + strand) is selectively replicated, packaged into phage coat proteins, and extruded from the cell. The M13 phage are released from the cell without lysis of the host. However, since reproduction of M13 slows the growth rate of the host, phage infected cells form a "turbid plaque" on a lawn of uninfected cells. A diagram of the life cycle of M13 is shown in Figure 10-1.

M13 AS A CLONING VECTOR

M13 is useful for obtaining single stranded DNA (ssDNA) for dideoxy sequencing (Smith, 1987). The RF is present at a high copy number so it can be easily purified. The double stranded DNA (dsDNA) insert can be cloned into restriction sites on the double stranded RF and the recombinant RF transfected into a competent host. (Transfection is simply transformation with viral DNA.) Since there are no structural constraints on the size of filamentous phage, the size of the insert DNA is theoretically unlimited. (In practice, phage with inserts greater than 5 kb are often unstable.) Since a high titer of phage is produced (about 10^{11} pfu/ml) and the phage particles are purified away from cells prior to extracting the DNA, it is easy to obtain very pure ssDNA that is not contaminated with cellular nucleic acids.

Messing and coworkers have constructed a number of M13 derivatives (M13mp) that are useful cloning vectors. The M13mp vectors have a small portion of the N-terminal sequence of *lacZ* cloned into the intergenic region of M13. A small polylinker is inserted in frame with the *lac* fragment. The polylinker has a number of useful restriction enzyme sites (Figure 10-2) and hence is called the multiple cloning site (MCS). There are no other sites for these enzymes on the phage, so the sites in the MCS are useful for cloning DNA fragments.

It is easy to screen for insertions into the MCS because most insertions into the MCS disrupt the *lacZ* peptide. In addition, primers that hybridize to the DNA adjacent to the MCS are available commercially, so the DNA cloned into the MCS can be easily sequenced.

Figure 10-1. Infection and growth of M13 phage

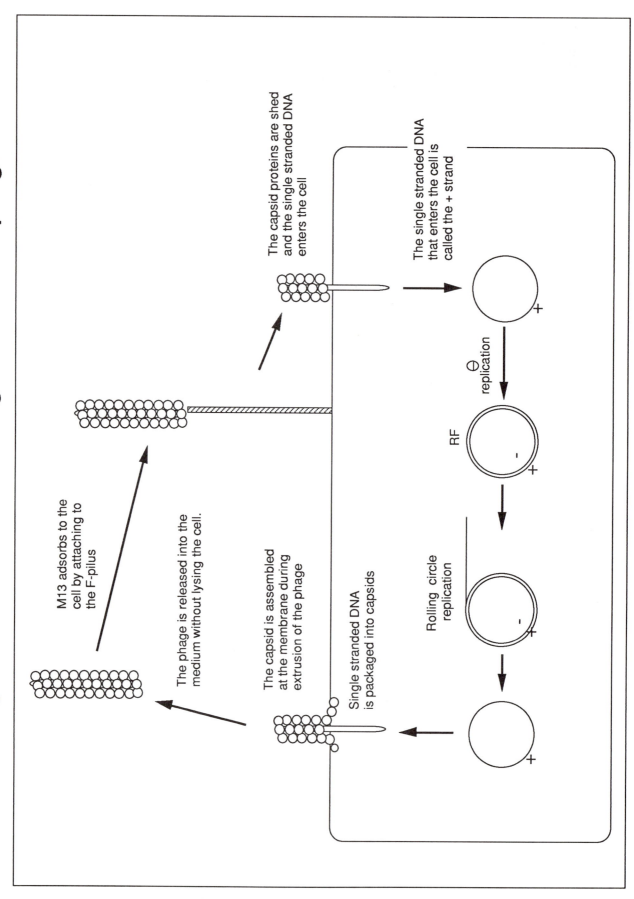

M13 adsorbs to the cell by attaching to the F-pilus

The phage is released into the medium without lysing the cell.

The capsid is assembled at the membrane during extrusion of the phage

Single stranded DNA is packaged into capsids

Rolling circle replication

The capsid proteins are shed and the single stranded DNA enters the cell

The single stranded DNA that enters the cell is called the + strand

replication

RF

+

−

+

−

+

+

The two M13 vectors we will use are mp18 and mp19. The two vectors are identical except the multiple cloning sites of mp18 and mp19 are in opposite orientations to facilitate cloning DNA fragments in both directions. A diagram of these vectors is shown in Figure 10-2.

LIGATION

In order to construct a recombinant clone, the DNA fragments must be covalently sealed. Enzymes that catalyze the formation of a phosphodiester bond between adjacent 3'OH and 5'P termini in double stranded DNA are called DNA ligases. *E. coli* and *S. typhimurium* have an NAD dependent DNA ligase that is essential for repairing nicks generated during DNA replication and recombination *in vivo*. The DNA ligase commonly used for DNA cloning is made by phage T4. In contrast to the enzyme from *E. coli*, T4 DNA ligase uses ATP instead of NAD. In addition, T4 DNA ligase can ligate blunt ends as well as sticky ends.

Several factors affect efficiency of ligation of vector and insert DNA *in vitro* (Rodriguez and Tait, 1983; Perbal, 1988).

(1) *Temperature.* Ligation of DNA fragments occurs when DNA ligase seals the phosphate bond between two DNA fragments that are very close to each other. DNA fragments with sticky ends can form hydrogen bonds, forming a "nicked" double stranded substrate for DNA ligase. The optimal temperature for T4 ligase is 37°C. However, at 37°C hydrogen bonds between the short sticky ends of most restriction sites are denatured, and T4 DNA ligase is not stable for long periods at 37°C. Therefore, ligation is usually done at a lower temperature that is a compromise between these opposing factors. Often 12-15°C is used for ligation of DNA fragments with short sticky ends. (Most ligations work at room temperature. Although the ligation efficiency is not quite as high at room temperature compared to lower temperatures, this eliminates the need for refrigerated water baths.) Since blunt ended fragments do not form hydrogen bonds, blunt end ligations work well at higher temperatures. Blunt end ligations depend upon the two DNA fragments and ligase simultaneously bumping into each other, therefore, higher concentrations of ligase are required for efficient blunt end ligations.

(2) *Concentration of ATP.* ATP is required by T4 DNA ligase, but too much ATP (>2.5 mM) inhibits ligation.

(3) *Concentration of DNA.* If the DNA concentration is too low the probability of two molecules bumping into each other is low, so the intermolecular ligation frequency will be low. Under these conditions, the ends of a DNA molecule are more likely to bump into each other than they are to bump into the end of another DNA molecule, so intramolecular ligation will be most common. If the DNA concentration is too high then more than two molecules may be ligated together, yielding multimers.

(4) *Ratio of vector and insert.* If the ratio of vector to insert is too high, vector molecules are more likely to bump into each other than the insert fragment, so ligation of vector multimers will be most common. If the ratio of vector to insert is too low, more than one insert may be ligated to each vector. (As a "rule of thumb," a ratio of 1 mole of vector to 3 moles of insert often works well.)

Figure 10-2. M13 phage vectors used.

Only restriction enzymes that cut the vectors at a single site are shown on the following restriction maps. The two vectors M13 mp18 and mp19 are identical except their multiple cloning sites are in opposite orientations to facilitate cloning DNA fragments in both directions. The multiple cloning sites of mp18 and mp19 are shown at the top of the figure. The amino acids in the α-peptide are shown with the amino acids encoded by the polylinker indicated in lower case letters. (Modified from the 1986 BRL catalog).

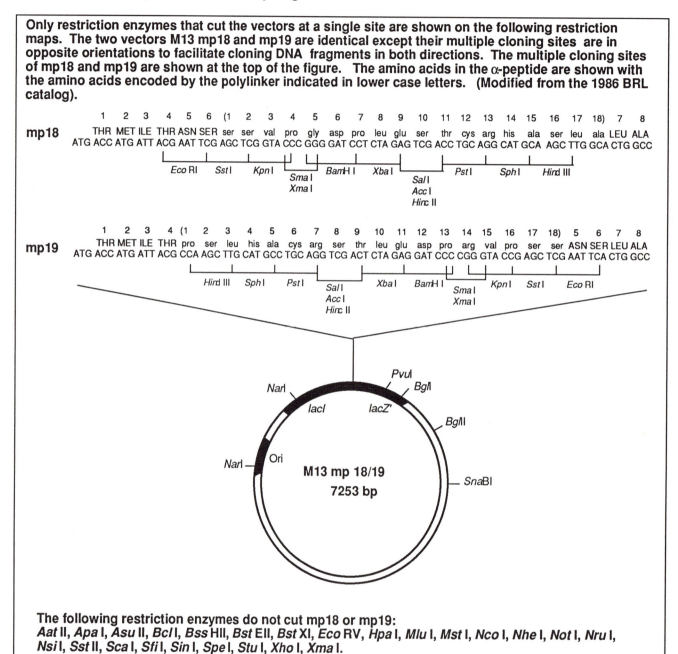

The following restriction enzymes do not cut mp18 or mp19:
Aat II, Apa I, Asu II, Bcl I, Bss HII, Bst EII, Bst XI, Eco RV, Hpa I, Mlu I, Mst I, Nco I, Nhe I, Not I, Nru I, Nsi I, Sst II, Sca I, Sfi I, Sin I, Spe I, Stu I, Xho I, Xma I.

SCREENING FOR CLONED INSERTS

After ligation and transfection into a suitable host, the presence of a cloned insert in the M13mp vector can be easily screened by checking β-galactosidase expression on Xgal plates. Since the M13mp vectors only contain a small portion of the *lacZ* gene, expression of β-galactosidase from M13mp requires α-complementation. The M13mp hosts carry an F′ with a portion of the *lacZ* gene [*lacZ* del(M15)] which lacks the N-terminus of β-galactosidase. The mutant protein produced by *lacZ* del(M15) has the active site of β-galactosidase but is inactive because it cannot tetramerize (Zabin and Fowler, 1980). (The active form of β-galactosidase is a tetramer.) However, if the short N–terminal β-galactosidase peptide encoded on the M13mp phage is provided in *trans* it can associate with the del(M15) mutant protein to form a functional β-galactosidase tetramer. The functional interaction of the two mutant polypeptides is called alpha-complementation (Figure 10-3).

β-galactosidase activity can be detected on plates containing the indicator Xgal (5-bromo-4-chloro-3-indoyl-β-D-galactoside). Xgal is colorless, but when cleaved by β-galactosidase an insoluble blue indigo dye is released. In order to express β-galactosidase, the lac operon must be induced. Xgal does not induce *lac* expression. Therefore, IPTG (isopropyl-β-D-thiogalactoside), a nonmetabolizable inducer of the *lac* operon, is included in the Xgal plates. When an appropriate host is infected, phage that carry the intact lac fragment form blue plaques on Xgal + IPTG plates but phage with cloned inserts in the MCS form colorless plaques.

References

Hackett, P., J. Fuchs, and J. Messing. 1984. *An Introduction to Recombinant DNA techniques*, pp. 34-44. Benjamin/Cummings Publishing Co., Menlo Park, CA.

Perbal, B. 1988. *A Practical Guide to Molecular Cloning*. Wiley-Interscience, NY.

Rodriguez, R., and R. Tait. 1983. *Recombinant DNA Techniques*, pp. 81-88. Benjamin/Cummings Publishing Co., Menlo Park, CA.

Smith, G. 1987. Filamentous phages as cloning vectors. *In* R. Rodriguez and D. Denhardt (eds.), *Vectors: A Survey of Molecular Cloning Vectors and Their Uses*, pp. 61-83. Butterworth and Co., Stoneham, MA.

Zabin, I., and A. Fowler. 1980. β-Galactosidase, the lactose permease, and thiogalactoside transacetylase. *In* J. Miller and W. Reznikoff (eds.), *The Operon*, pp. 89-121. Cold Spring Harbor Laboratory, NY.

Zinder, N., and K. Horiuchi. 1985. Multiregulatory element of filamentous bacteriophages. *Microbiol. Rev. 40*:101-106.

Figure 10-3. Alpha-complementation

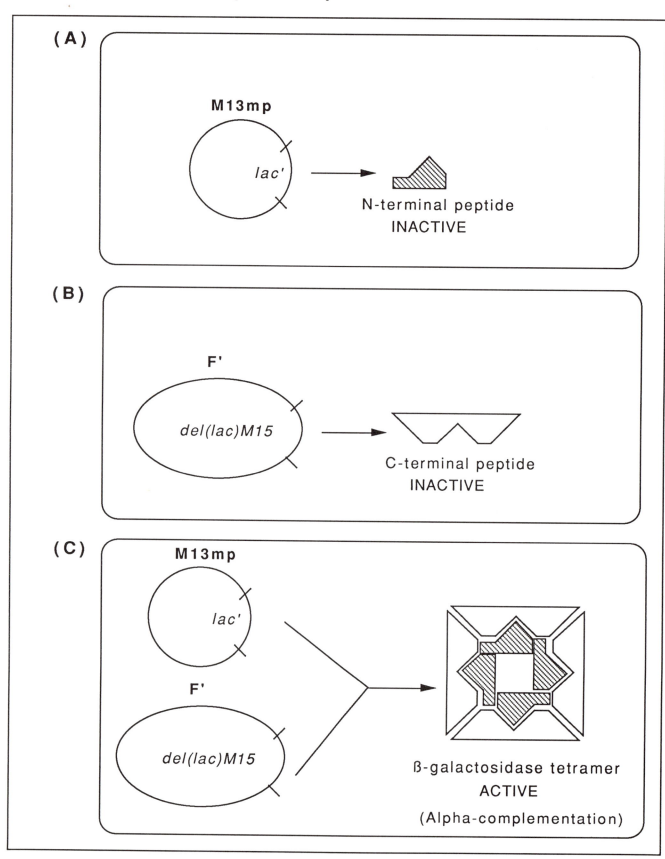

10A. Purification of Restriction Fragments

DNA fragments from the cloned gene could simply be digested with a restriction enzyme and the restriction fragments directly subcloned into a M13 vector cut with the same enzyme. However, cloning and screening for a desired insert is often tedious and time consuming. Therefore, in order to subclone a specific restriction fragment into M13 for DNA sequence analysis, the desired fragment can be purified from an agarose gel. Two techniques for purification of restriction fragments from agarose gels are described: phenol extraction of frozen bands cut out of agarose gels and electroelution from agarose gels. Both methods require initial separation of the DNA fragments on an agarose gel. (Many other techniques for purifying restriction fragments are available, but the two techniques described here are simple, inexpensive, and result in a high yield of DNA.)

1.   Digest the plasmid DNA in a sterile microfuge tube as follows:

Sterile dH$_2$0	__ µl
DNA (1-2 µg)	__ µl
10x RE buffer	5 µl
RE (1-2 units)	1 µl
Total	50 µl

The restriction enzyme used will depend on the fragment to be subcloned. Appendix 6A shows the site specificity of some useful restriction enzymes and the appropriate reaction buffers.

2.   Incubate digests for 1-2 hrs at 37°C. Verify complete digestion by running a small aliquot of the digested DNA and undigested DNA on an agarose minigel (Appendix 7A). Stain the gel with ethidium bromide and examine it on the transilluminator (Appendix 7D). WEAR GLOVES. If digestion is not complete, continue incubation at 37°C.

3.   Add 5 µl of Blue II and run the entire sample on a 0.8% agarose gel overnight (Appendix 7A). Run Lambda HindIII standards as size markers (1 µl Lambda HindIII + 8 µl TE + 1 µl Blue II).

4.   Stain the gel with about 0.5 µg/ml ethidium bromide for about 30 min. WEAR GLOVES. Keep the gel covered during staining.

5.   Place the ethidium bromide stained agarose gel on a glass plate. Illuminate from above with UV. WEAR EYE PROTECTION. Cut the bands of interest out of the agarose gel with a spatula. The gel slices can be stored in microfuge tubes at 4°C if necessary.

Method 1: Phenol extraction from frozen agarose

CAUTION: Wear gloves to avoid contact with ethidium bromide from the agarose gel and to avoid contaminating the DNA fragments with nucleases from your skin flora.

1. Freeze the gel fragments in a dry-ice ethanol bath (about 10 min). The frozen bands can be stored at -20°C if necessary.
2. Thaw the fragments at room temperature. While still partially frozen, thoroughly mash the fragments. (A make-shift pestle can be constructed by passing a disposable plastic pipetman "tip" through a flame to seal the end.)
3. Repeat steps 1 and 2 two more times.
4. Add 100 µl TE saturated phenol. Vortex 10 sec.
5. Freeze 10 min in a dry-ice ethanol bath.
6. Spin 15 min at room temperature in a microfuge.
7. Remove the upper aqueous phase with a pipetman and place in a clean microfuge tube. Add 100 µl TE to the phenol phase and reextract. Combine the TE phases.
8. Reextract the TE phases twice with an equal volume of TE saturated phenol.
9. Remove the upper aqueous phase and place in a clean microfuge tube. Ethanol precipitate the DNA by adding a half volume of 7.5 M ammonium acetate and 2.5x the total volume of 95% ethanol. Place on ice for 30 min.
10. Spin 15 min in the microfuge.
11. Pour off the supernatant. Overlay the pellet with 1 ml of cold 70% ethanol. Spin 5 min in the microfuge.
12. Pour off the supernatant. Drain the tube over a Kimwipe. Dry 15-20 min in a vacuum dessicator to remove any residual ethanol.
13. Resuspend the pellet in 10 µl ddH$_2$0. Run 1 µl on a 0.8% agarose minigel (Appendix 7A) to determine the yield of purified DNA (use 1 µl sample + 9 µl TE + 1 µl Blue II). Stain the gel with ethidium bromide and photograph it (Appendix 7D).

Reference

Benson, S. 1984. A rapid procedure for isolation of DNA fragments from agarose gels. *Biotechniques* 2: 66-67.

Method 2: Electroelution from agarose gels

1. Prepare a gel tray without the comb. Prepare 100 ml of 0.8% agarose in 1x TBE. Add ethidium bromide to 5 µg/ml.
2. Place the cut out gel slices in the tray parallel to the original band.
3. Cool the agarose solution to about 55°C then slowly pour it into the gel tray being careful to avoid disturbing the gel slice. The agarose should be level with the top of the gel slice.
4. After the agarose solidifies, cut a V about 2.5 mm in front of the gel slice (toward the + electrode). Cut completely through the agarose gel with a spatula (be careful to avoid damaging the plastic tray).
5. Cut a piece of dialysis membrane and a piece of 3MM paper slightly longer than the DNA band and wide enough to stick up above the gel about 1 mm.
6. Dampen the 3MM paper in 1x TBE. Place the dialysis membrane on the dampened 3MM paper. Using forceps, slide the dialysis tubing and 3MM paper into the V cut in the gel with the 3MM paper toward the DNA (see below).

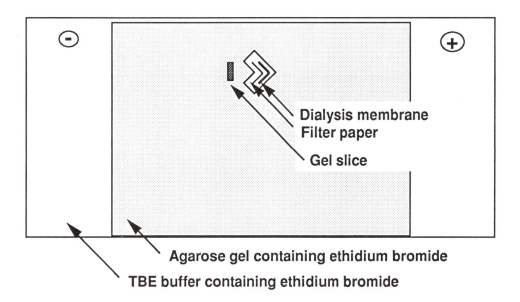

Dialysis membrane
Filter paper
Gel slice

Agarose gel containing ethidium bromide
TBE buffer containing ethidium bromide

7. Add enough 1x TBE buffer containing 5 µg/ml ethidium bromide to just submerge the gel. (The 3MM paper should stick up above the level of the buffer).
8. Run the gel at 100-150 volts until the DNA has completely migrated onto the paper. Monitor the migration of the DNA with a hand held UV lamp.
9. Remove the dialysis tubing/3MM paper "sandwich" with forceps. Place in a small microfuge tube that has a small hole poked in the bottom. (If you heat an 18 guage needle in a flame, the hot needle will easily poke a hole in a microfuge tube. Be careful to avoid poking yourself with the hot needle.)
10. Place the small tube inside a sterile large microfuge tube and spin for 15 sec in the microfuge.
11. Transfer the eluate to a clean sterile large microfuge tube. Wash the dialysis tubing/3MM paper by adding 50 µl electroelution buffer to the small microfuge tube and spinning for 15 sec.

12. Combine the wash with the first eluate. Repeat step 11.
13. Extract the eluate twice with 150 µl phenol:chloroform:isoamyl alcohol (25:24:1).
14. Extract once with chloroform:isoamyl alcohol (24:1).
15. Ethanol precipitate the DNA. Add 1/2 volume of 7.5 M ammonium acetate and 2.5x the total volume of 95% ethanol. Place in ice-water bath for 30 min.
16. Spin 15 min in a microfuge.
17. Dry 15-20 min in vacuum dessicator to remove any residual ethanol.
18. Resuspend the pellet in 10 µl ddH$_2$0. Run 1 µl on a 0.8% agarose minigel (Appendix 7A) to determine the yield of purified DNA (1 µl sample + 9 µl TE + 1 µl Blue II). Stain the gel with ethidium bromide and photograph it (Appendix 7D).

Reagents

Electroelution buffer (200 mM NaCl, 50 mM Tris, 1 mM EDTA, 0.1% SDS)

400 µl	0.5 M NaCl
500 µl	1 M TrisHCl (pH 8)
20 µl	0.5 M EDTA
100 µl	10% SDS
10 ml	ddH$_2$O

Reference

Maniatis, T., E. Fritsch, and J. Sambrook. 1980. *Molecular Cloning: A Laboratory Manual*, pp. 168-169. Cold Spring Harbor Laboratory, NY.

10B. Restriction Digests of M13 RF

1. Digest the M13 RF DNA in a sterile microfuge tube as follows:

Sterile ddH$_2$0	__ μl
DNA (1-2 μg)	__ μl
10x RE buffer	1 μl
RE	1 μl
Total	10 μl

The restriction enzyme used will depend on the fragment purified from the plasmid clone. Appendix 6A shows the site specificity of restriction enzymes and the appropriate restriction enzyme buffer.

2. Incubate digests for 1-2 hrs at 37°C. Verify complete digestion by running a small aliquot on a 0.8% agarose minigel (Appendix 7A). If digestion is not complete, continue incubation at 37°C.

3. Add an equal volume of chloroform:isoamyl alcohol (24:1) and vortex. Spin for 1 min in the microfuge.

4. Remove the top aqueous layer. Combine the digested M13 RF with the purified plasmid restriction fragment in a microfuge tube at a molar ratio of about 1:3. As a control add an equal volume of the digested M13 RF without insert to another microfuge tube. Save any remaining DNA in the refrigerator.

5. Ethanol precipitate the DNA as follows:

    a. Add 1/2 volume of 7.5 M ammonium acetate.

    b. Add 2.5x the total volume of 95% ethanol.

    c. Mix. Place on ice for 30 min.

    d. Spin 15 min in a microfuge.

    e. Pour off the ethanol. Drain the tube over a Kimwipe.

    f. Place the microfuge tube with the DNA pellet in a vacuum dessicator for 15-20 min to dry off any residual ethanol.

6. Resuspend the pellet in 18 μl of sterile ddH$_2$0.

7. Remove 1 μl and save in a microfuge tube in the refrigerator for a preligation control.

10C. Ligation

1. Add 2 μl of 10x Ligase buffer and 1 μl of T4 DNA ligase (5-10 units). Mix with a pipetman.
2. Incubate at room temperature overnight.
3. Remove 1 μl as a postligation control.
4. Add 9 μl TE and 1 μl Blue II to the pre- and post-ligation controls and run an agarose minigel with Lambda *Hin* dIII standards (Appendix 7A). Stain with 0.5 μg/ml ethidium bromide and photograph (Appendix 7D).

Reagents

10x Ligase buffer

      0.5 ml 1 M TrisHCl pH 7.6
      0.1 ml 1 M MgCl$_2$
      0.1 ml 100 mM DTT
      0.1 ml 10 mg/ml BSA (Bovine serum albumin, Pentax fraction V)
      0.1 ml 100 mM ATP
      0.1 ml sterile ddH$_2$0
      Store frozen in small aliquots at -20°C.

0.1 M ATP

      60 mg ATP
      800 μl ddH$_2$0
      Neutralize to pH 7 with 0.1 M NaOH. (NTPs undergo acid catalyzed hydrolysis if they are not neutralized.)
      Bring to 1 ml with ddH$_2$0.
      Store frozen in 0.1 ml aliquots at -20°C.

10D. Transfection with M13

1. Grow EM383 [*E. coli hsd-5* del(*lac-proAB*) *supE thi* /F'ProA⁺B⁺ *lacI*q *lacZ* del(M15)] on E + glucose + thiamine plates without proline to prevent segration of the F'.
2. Grow EM383 overnight in 20 ml LB at 37°C. (Enough for entire class.)
3. Subculture 0.2 ml into 20 ml LB in a Klett flask. (Save the rest of the cells for step #13.) Grow to about 50 Klett units on a 37°C shaker.
4. Cool cells on ice 20-30 min.
5. Spin down cells 5 min at 5000 rpm in SS34 rotor at 4°C.
6. Gently resuspend pellet in 10 ml ice cold 50 mM $CaCl_2$ DO NOT VORTEX. Keep on ice for about 30 min.
7. Spin down cells 5 min at 5000 rpm in SS34 rotor at 4°C.
8. Gently resuspend cells in 1.0 ml ice cold 50 mM $CaCl_2$. DO NOT VORTEX. Keep on ice at least 30 min. The cells are competent for transformation after this step.
9. Bring the ligated DNA to 17 µl with TE. In glass test tubes on ice add aliquots of DNA and competent cells as shown below:

Tube	DNA	Competent cells	
A	—	0.1 ml	Cell control
B	1 µl ligation mix	0.1 ml	
C	3 µl ligation mix	0.1 ml	
D	6 µl ligation mix	0.1 ml	
E	6 µl ligation mix	—	DNA control

10. Keep on ice for 30 min.
11. Place in a 37°C water bath for 2 min to heat pulse then place on ice for about 30 min.
12. Add 0.2 ml LB and place in 37°C shaker for about 30 min.
13. Melt TS top agar in the microwave. Add 2.5 ml top agar to 5 sterile test tubes and place in a heating block at about 50°C. It is important to allow the top agar to cool to about 50°C before adding the cells.
14. Just before use add 50 µl Xgal + 50 µl IPTG + 0.1 ml fresh EM383 to each tube of top agar to make "complete" top agar.
15. Pour one tube of the complete top agar into one of the transformation tubes. Quickly swirl to mix, pour on a fresh NB plate, and rock the plate so the top agar covers the entire surface. Repeat with each tube.
16. Allow the top agar to solidify for about 10 min on the bench.
17. Incubate the plates upside-down at 37°C overnight. Start a fresh culture of EM383 in 2 ml LB and grow overnight at 37°C.
18. Count the number of blue and colorless plaques obtained. Pick the colorless plaques by removing an agar plug with a Pasteur pipet and blowing the agar plug into 1 ml of sterile 0.85% NaCl. Vortex.
19. Plaque purify the phage as follows:
    a. Dilute each phage suspension to 10^{-6}, 10^{-7}, and 10^{-8} in sterile 0.85% NaCl.
    b. Mix 0.1 ml of each phage dilution with 0.1 ml fresh EM383 in sterile tubes.
    c. Add 2.5 ml TS top agar + 50 µl Xgal + 50 µl IPTG to each tube.
    d. Immediately pour onto an LB plate and gently rock the plate so the top agar covers the entire surface.
    e. Allow to solidify for about 10 min then incubate upside-down at 37°C

overnight.

NOTE. It is also possible to purify M13 plaques by streaking the phage suspension on a LB plate then overlaying with TS top agar containing Xgal + IPTG + EM383.

Reagents

50 mM CaCl$_2$

        1.5 ml sterile 1 M CaCl$_2$

        28.5 ml sterile ddH$_2$0

        Cool on ice before use.

Xgal (5-Bromo-4-chloro-3-indoyl-ß-D-galactoside)

        20 mg Xgal dissolved in 1 ml N,N-dimethylformamide.

IPTG (Isopropyl-ß-D-thiogalactoside)

        5 mg IPTG dissolved in 1 ml sterile dH$_2$0.

RESULTS
EXPERIMENT 10

1. Show a restriction map of the plasmid indicating which fragment was purified.
2. Include photocopies of all agarose gels (before and after pictures of the purified DNA fragment, pre- and post-ligation controls).
3. Include a table showing the number of blue and colorless M13 plaques obtained on each plate.

11

DNA SEQUENCE ANALYSIS

DIDEOXY SEQUENCING

The purpose of this experiment is to determine the sequence of the DNA cloned into M13 vectors in Experiment 10.

Dideoxy sequence analysis is based upon the random incorporation of analogs of the deoxynucleoside triphosphates (dNTPs) into a growing DNA chain by DNA polymerase. These dideoxynucleoside triphosphate (ddNTP) analogs lack the 3'-OH group on the ribose moiety of the nucleotide.

dNTP

ddNTP

The 3'-OH group is necessary for the formation of the next phosphodiester bond so incorporation of a ddNTP into a growing DNA chain causes termination of chain elongation. To determine the sequence of a DNA template, four separate reactions are run. Each reaction contains all four dNTPs but only one of the four ddNTPs. For every nucleotide on the template, DNA polymerase inserts the complementary nucleotide during synthesis of the new DNA strand. If a dNTP is inserted chain elongation continues, but if a ddNTP is inserted synthesis stops at that position. For example, in the reaction with ddATP, when the enzyme needs to incorporate an adenine nucleotide it has the choice between the substrates dATP and ddATP. If it incorporates the ddATP then the reaction stops (chain termination). If it incorporates the dATP the reaction continues until another adenine nucleotide is needed, then the enzyme again has a choice between the dATP or ddATP. By controlling the ratio of ddATP to dATP in the reaction, incorporation of the ddATP will be random. This results in a nested set of DNA fragments of different lengths, each terminated at a different adenine residue. By determining the length of fragments produced with each of the four ddNTPs it is possible to deduce the nucleotide sequence of the template DNA (see Figure 11-1).

Initiation of DNA synthesis requires double stranded DNA as a primer. The primer is provided by using a small oligonucleotide which hybridizes to the single stranded DNA adjacent to the multiple cloning site. DNA polymerase begins synthesis from the 3' end of the primer and adds dNTPs in the 5' to 3' direction through the multiple cloning site into the cloned DNA (Figure 11-1). Thus, every DNA fragment has the same 5' end but different 3' ends (wherever a ddNTP was inserted). The Klenow fragment of DNA polymerase is used for dideoxy sequencing because it lacks the 5' to 3' exonuclease activity present in the *E. coli* DNA polymerase I holoenzyme. This exonuclease activity would degrade the common 5' end of the DNA fragments making interpretation of the DNA sequence impossible.

By including a radioactive dNTP (e.g., α-^{32}P-dATP or α-^{35}S-dATP), the DNA fragments become radioactively labeled during synthesis. After the reactions are stopped, the sequencing fragments are denatured from the template and resolved according to size by polyacrylamide gel electrophoresis. If the DNA fragments are denatured, fragments that differ in length by a single nucleotide can be separated on polyacrylamide gels. The gels contain urea and are run at high voltages (which makes them hot) to keep the DNA denatured. The DNA sequence is determined following autoradiography of the gel by reading the order of the bands in the four lanes from each ddNTP reaction. The bands form a ladder corresponding to the size of the DNA fragments. The first band at the bottom of the gel represents the shortest fragment synthesized from the sequencing primer that terminated with the ddNTP used. (Usually the smallest readable band is 5-10 bp from the end of the primer). The sequence is determined by reading up the four lanes of the autoradiogram in order of the occurrence of the bands on the ladder.

NOTE: DNA sequencing kits containing all the necessary solutions are available from many companies. I highly recommend a DNA sequencing kit sold by US Biochemicals that uses modified T7 DNA polymerase ("Sequenase"): it is almost foolproof, relatively inexpensive, and gives beautiful results. If a DNA sequencing kit is used, follow the directions for the DNA sequencing reactions supplied by the manufacturer.

References

Bankier, A. 1984. Advances in dideoxy DNA sequencing. *Biotechniques* 2: 72-77.

Kornberg, A. 1980. *DNA Replication*, pp. 479-496. W.H. Freeman and Company, San Francisco, CA.

Sanger, F., S. Micklen, and A. Coulson. 1977. DNA sequencing with chain-terminating inhibitors. *Proc. Natl. Acad. Sci.* 74: 5463-5467.

Figure 11-1. Dideoxy DNA sequencing

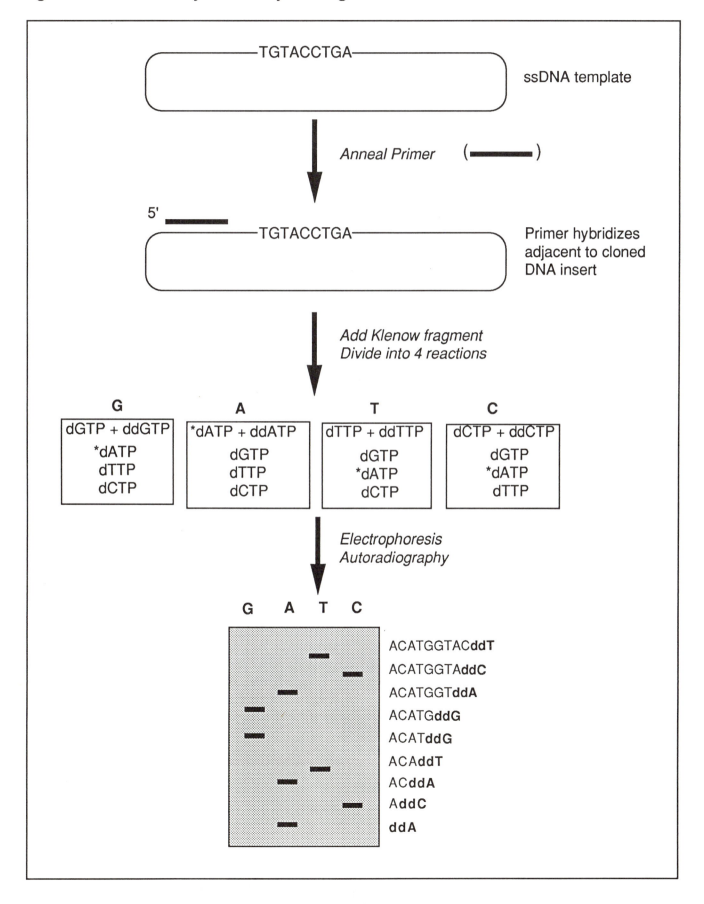

11A. Isolation of Single Stranded DNA (ssDNA)

1. Add 50 µl fresh EM383 and a freshly "cored" plaque of M13 to 5 ml LB in a small flask. Grow with vigorous shaking at 37°C for about 5 hr.
2. PEG precipitation of phage:
   a. Transfer 1.4 ml of the M13 culture to a 1.5 ml microfuge tube.
   b. Spin 5 min in a microfuge.
   c. Transfer the supernatant to a clean microfuge tube. Be careful to avoid the cell pellet.
   d. Add 200 µl PEG/NaCl. Incubate on ice for 5 min.
   e. Spin 5 min in a microfuge.
   f. Aspirate off the supernatant. (A simple aspiration setup can be made with a water aspirator attached to a vacuum hose to a Pasteur pipet with a pipetman tip on the end.)
   g. Respin 30 sec. Aspirate off any residual PEG.
   h. Resuspend in 300 µl TE. (If necessary, the phage can be stored overnight at 4°C after this step).
3. Phenol extraction of phage DNA:
   a. WEAR GLOVES. Add 300 µl TE saturated phenol. Tightly cap the microfuge tube and vortex 10 sec. Spin 2 min in a microfuge.
   b. Remove upper aqueous layer and place in a clean microfuge tube. Reextract with TE saturated phenol two more times.
   c. In the fume hood add 0.5 ml ethyl ether. **CAUTION - ETHER IS VERY FLAMMABLE. ONLY USE ETHER IN A FUME HOOD AND MAKE SURE THERE ARE NO FLAMES NEARBY.** Vortex.
   d. Aspirate off the upper ether layer in the fume hood. Repeat the ether extraction two more times.
   e. Evaporate off residual ether on a 60°C heating block.
   f. Add 1/10 volume 3 M sodium acetate and 2.5 volumes of 95% ethanol.
   g. Place on ice for 30 min to precipitate the DNA.
   h. Spin 10 min in a microfuge.
   i. Pour off the supernatant. Add 0.5 ml 70% ethanol and respin.
   j. Pour off the supernatant. Drain over a Kimwipe. Dry off residual ethanol for 15-20 min in a vacuum dessicator.
   k. Resuspend in 20 µl TE. Run 2 µl on an agarose minigel (2 µl DNA + 8 µl TE + 1 µl Blue II) and stain with ethidium bromide (Appendix 7) to check your yield of ssDNA. Store the rest at -20°C.

11B. Preparation of Sequencing Gels

1. Clean the gel plates thoroughly with a non-abrasive detergent. Rinse thoroughly with dH_2O, and dry with a Kimwipe.
2. Thoroughly rinse the plates with ethanol and wipe with Kimwipes. Avoid touching the plates with your fingers.
3. Assemble the plates with the clean surface facing inside. Insert the side spacers. Tape the sides and bottom of the plates and clamp the sides.
4. Prepare the gel as described below.
   a. Add the following reagents to a small vacuum flask:
50 g	Ultra-pure urea
20 ml	acrylamide/bis-acrylamide (38:2)
10 ml	10x TBE
30 ml	ddH_2O
   b. Warm to room temperature to completely dissolve the urea.
   c. Filter through Whatman #1 paper.
   d. Cover the top of the flask with a stopper, attach the arm to a vacuum, and gently swirl to degas.
5. Add 1 ml of ammonium persulfate and 30 µl TEMED. Swirl gently to mix.
6. Immediately pour the gel. Hold the plate at a 45° angle. Pour the acrylamide/ urea solution down one side of the plate in a single, smooth motion to avoid air bubbles (see Figure 11-2). After the plates are full, lay the gel down flat on a table.
7. Insert the flat side of a sharks-tooth comb. Clamp the plates together over the comb.
8. Allow the gel to polymerize for at least 1 hr. Save a little of the acrylamide solution in a Pasteur pipet to check for polymerization.
9. After polymerization is complete, remove the clamps and tape. Carefully remove the comb. Rinse the top of the gel with 1x TBE to remove any unpolymerized acrylamide. Rinse the comb with dH_2O to clean off any acrylamide. Invert the comb and insert between the plates so the teeth just touch the gel surface. Do not puncture the gel surface or move the comb once it has been inserted.
10. Clamp the gel onto the electrophoresis apparatus. Pour 1x TBE buffer into the top and bottom buffer chambers. Flush out the wells with 1x TBE to remove any unpolymerized acrylamide.

Figure 11-2. DNA sequencing gels

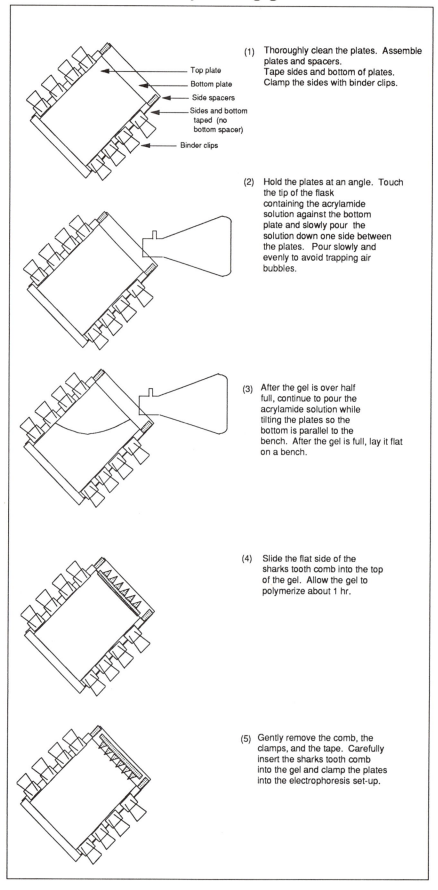

(1) Thoroughly clean the plates. Assemble plates and spacers.
Tape sides and bottom of plates.
Clamp the sides with binder clips.

Top plate
Bottom plate
Side spacers
Sides and bottom taped (no bottom spacer)
Binder clips

(2) Hold the plates at an angle. Touch the tip of the flask containing the acrylamide solution against the bottom plate and slowly pour the solution down one side between the plates. Pour slowly and evenly to avoid trapping air bubbles.

(3) After the gel is over half full, continue to pour the acrylamide solution while tilting the plates so the bottom is parallel to the bench. After the gel is full, lay it flat on a bench.

(4) Slide the flat side of the sharks tooth comb into the top of the gel. Allow the gel to polymerize about 1 hr.

(5) Gently remove the comb, the clamps, and the tape. Carefully insert the sharks tooth comb into the gel and clamp the plates into the electrophoresis set-up.

Reagents

Acrylamide/bis-acrylamide stock (38:2)
CAUTION - ACRYLAMIDE IS TOXIC. AVOID CONTACT WITH SKIN.
> Dissolve 76 g Ultra-pure acrylamide and 4 g bis-acrylamide in about 150 ml ddH$_2$0.
> Bring to 200 ml with ddH$_2$0.
> Filter through Whatman #1 paper.
> Store in a brown bottle at 4°C.

Ammonium persulfate (100 mg/ml)
> Dissolve 1 g ammonium persulfate in 9 ml ddH$_2$0.
> Aliquot and store at -20°C.

10x TBE
> 162   g   Tris base
> 27.2 g   Boric acid
> 9.3 g   Na$_2$EDTA
> Dissolve in about 600 ml ddH$_2$0 on a magnetic stirrer.
> Bring to 1 liter with ddH$_2$0. Filter through Whatman #1 filter paper.
> Store at room temperature. Discard if a solid precipitate develops.

11C. Dideoxy Sequencing Reactions

CAUTION: This experiment uses radioactive materials. WEAR GLOVES AND A LAB COAT.

1. Preheat a heating block containing water-filled test tubes to about 90°C.
2. In addition to your purified ssDNA use M13 ssDNA as a control. Add the following reagents to two small microfuge tubes:
   - 1.0 µl   10x TMS buffer
   - 2.0 µl   Primer
   - 6.4 µl   ddH$_2$O
   - 3.0 µl   ssDNA (1 mg/ml)
3. Place the microfuge tubes in the water-filled test tubes in the heating block for 5 min. (This denatures secondary structures in the ssDNA.)
4. Remove the glass test tubes containing the microfuge tubes and allow to slowly cool to room temperature (about 30 min). (Leaving the microfuge tube in the water-filled test tube slows the rate of cooling to allow efficient annealing of the primer to the ssDNA template.)
5. While the template and primer are annealing, label four small microfuge tubes A, C, G, and T for each ssDNA template. Add the following reagents to the sides of the tubes:
   - A: 1 µl A sequencing mix + 1 µl diluted ddATP
   - C: 1 µl C sequencing mix + 1 µl diluted ddCTP
   - G: 1 µl G sequencing mix + 1 µl diluted ddGTP
   - T: 1 µl T sequencing mix + 1 µl diluted ddTTP
*6. After cooling, add to the hybridization mixture:
   - 1 µl   0.1 M DTT (diluted in TMS buffer)
   - 1 µl   α-^{35}S-dATP
   - 1 µl   Klenow fragment (diluted to 1 unit/µl in TMS)
*7. Add 3 µl of the hybridization mixture to the side of each tube.
*8. Spin 15 sec in a microfuge to mix the reagents.
*9. Incubate at 30°C for 15 min.
*10. Add 1 µl of the dNTP Chase mix to each tube.
*11 Incubate at 30°C for 15 min.
*12. Add 10 µl dye/formamide mix and vortex well.
*13. Heat the samples for 3 min at 90°C and load 2-3 µl per lane.
*14. Carefully monitor the area for radiation spills.

Reagents

α-^{35}S-dATP (10 mCi/ml, >800 Ci / mmol)

10x TMS buffer
   - 0.50 ml   1 M TrisHCl pH 8
   - 0.25 ml   1 M MgCl$_2$
   - 0.33 ml   5 M NaCl
   - 3.95 ml   ddH$_2$O

NTP dilution buffer
   - 0.2 ml   1 M KPO$_4$ (pH 7.0)
   - 4.0 ml   5 M NaCl
   - 95.8 ml   ddH$_2$O

10 mM stock solutions of dNTPs and ddNTPs:
  a.   Weigh out the appropriate amount of NTP to make 1 ml of a 10 mM stock (see following Table). Place in a sterile 5 ml plastic tube.
  b.   Add 1 ml sterile ddH$_2$O. Mix to dissolve.
  c.   Dilute 5 µl of each stock into 1 ml of NTP dilution buffer for OD readings.
  d.   Measure absorbance vs a buffer blank at the maximal absorbance for each specific NTP (see Table below).
  e.   Often the actual concentration is different from the expected concentration. The actual concentration of each stock solution can be calculated using the following formula:

$$c = A / \varepsilon l$$

  Where:
      c = concentration in moles/liter (M)
      ε = extinction coefficient in (M^{-1} cm^{-1})
      l = pathlength (width of cuvette = 1 cm)

  f.   Note the corrected concentration on the undiluted stock and store at -20°C. These mixes are good for months.

Physical constants of dNTPs and ddNTPs:

NTP	Molecular Weight	mg/ml for 10 mM Stock	Lambda Max (nm)	ε (1/moles cm)
dATP	589.2	5.9	259	15.2 x 10^3
dTTP	583.1	5.8	267	9.6 x 10^3
dCTP	569.0	5.7	271	13.1 x 10^3
dGTP	609.0	6.1	253	13.7 x 10^3
dITP	612.2	6.1	249	11.8 x 10^3
ddATP	617.0	6.2	259	15.0 x 10^3
ddTTP	608.2	6.1	167	9.6 x 10^3
ddCTP	542.0	5.4	280	13.0 x 10^3
ddGTP	615.2	6.2	253	13.7 x 10^3

Reference: 1985 Pharmacia P-L Biochemicals Catalog

DNA Sequencing mixes:

	A mix	C mix	G mix	T mix
µl 0.5 mM dCTP	20	1	20	20
µl 0.5 mM dGTP	20	20	1	20
µl 0.5 mM dTTP	20	20	20	1
µl 10x TMS	20	20	20	20

When stored at -20°C these mixes are good for several months.

ddNTP solutions

The concentration of ddNTPs used in the sequencing reaction depends upon the distance of the desired sequence from the primer and the type of isotope used in the sequencing reactions. The ddNTP concentrations we currently use for sequencing with ^{35}S-dATP are:

0.10 mM ddATP (1:100 dilution of 10 mM stock solution)
0.25 mM ddCTP (1:40 dilution of 10 mM stock solution)
0.25 mM ddGTP (1:40 dilution of 10 mM stock solution)
1.00 mM ddTTP (1:10 dilution of 10 mM stock solution)

dNTP Chase mix (1 mM of each dNTP)
1 µl   10 mM dATP
1 µl   10 mM dCTP
1 µl   10 mM dGTP
1 µl   10 mM dTTP
6 µl   ddH$_2$O

Dye/Formamide mix
3.2 ml   Deionized formamide
200 µl   10x TBE
400 µl   1% Xylene cyanol + 1% Bromphenol blue
200 µl   ddH$_2$O

11D. Electrophoresis and Autoradiography

1. Pre-run the gel for 30-60 min at 40 W constant power to heat up the gel.
*2. Heat the DNA sequencing reactions at 95°C for 3 min, then carefully load 2-3 μl of each denatured DNA sequencing reaction per lane using a pipetman. Keep track of the order in which the samples are loaded on the gel.
3. Electrophorese at 40 W constant power until the bromphenol blue dye reaches the bottom of the gel (2-3 hrs).
*4. Turn off the power and unplug the power supply. Remove the gel from the sequencing set-up and lay on the bench with the large plate facing up. Remove the tape from the plates. Carefully separate the plates by gently prying between them with a spatula. (Note which plate the gel sticks to so that you can orient your gel later).
*5. Place the gel in a tray and gently cover with 10% methanol + 10% acetic acid. Soak for about 10 min. Carefully aspirate off the methanol/acetic acid into a radioactive waste bottle. Gently cover the gel with dH$_2$O, soak for about 10 min, then aspirate off the dH$_2$O as above.
*6. Slowly lay a sheet of Whatman filter paper over the gel. Smooth out the air bubbles so the paper contacts the entire gel.
*7. Gently peel the filter paper off the plate. The gel sticks to the paper.
*8. Prepare a cold trap between the vacuum pump and the gel dryer. Cover the top of the gel with plastic wrap and dry on the gel dryer at 80°C for about 1 hr.
*9. Remove the plastic wrap and place the dried gel in a film holder. In the dark room, turn off the lights, remove a sheet of X-ray film, and immediately replace the cover on the box of film. Lay the sheet of X-ray film over the gel, then close the film holder, making sure it is "light tight". Expose the film 48-72 hrs at room temperature.
*10. Develop the film as described below:
    a. Remove the film in the dark room with the lights turned off and the safelight turned on.
    b. Submerge the film in Kodak X-ray developer and agitate intermittently for about 2 min.
    c. Rinse the film in water for about 30 sec.
    d. Fix the film for 5 min in Kodak rapid fixer.
    e. Rinse the film in running water for at least 15 min.
    f. Hang the film by the edge to air dry.

11E. Reading DNA Sequencing Audiograms

When reading a gel always read the spaces as well as the bands. Be sure that there are no blank spaces in the autoradiogram. When there are runs of bases, often not all the bases will have the same intensity. The following rules apply for doublets:

- The upper C is more intense than the lower C
- The upper G is often more intense than the lower G (especially when they follow a T)
- The upper A is often less intense than the lower A

The table on the following two pages lists some potential problems and solutions.

References

Blakesley, R. 1983. Diagnostic dideoxy DNA sequencing. *BRL Focus 5*: 1-4.

Ornstein, D., and M. Kashdan. 1985. Sequencing DNA using ^{35}S-labeling: a troubleshooting guide. *Biotechniques 3*: 476-483.

PROBLEM	POSSIBLE CAUSES	SOLUTION
Bands in all four lanes		
at high molecular weight (>50 bp)	Secondary structure of template DNA	Run the reaction at a higher temperature
at low molecular weight (25-45 bp)	Specific activity of dATP too high; low [dATP] limited elongation	Try lower spe-activity (400-800 Ci/mmol) dATP
at all molecular weights	Poor quality DNA polymerase	Use a new batch of DNA polymerase
	DNA polymerase lost activity during storage	Store at -20°C; do not dilute until just before use
	Template DNA contained impurities	Repurify ssDNA
	Primer was degraded	Use fresh primer
	Second site priming	Use a longer primer
Lack of large fragments	ddNTP:dNTP ratio too high	Reduce the [ddNTP]
Lack of small fragments	ddNTP:dNTP ratio too low	Increase the [ddNTP]
Autoradiogram is too light	Poor annealing of primer to template	Increase denaturing temperature; allow to reanneal more slowly
	Film developer exhausted	Prepare new developer
	Insufficient exposure	Increase the exposure time or use intensifying screens
High background in all four lanes	Template DNA not pure	Repurify template
	^{32}P-labeled DNA too old	Prepare fresh ^{32}P sequencing reactions or use ^{35}S-dNTP
Blurred bands in some parts of the autoradiogram	Poor contact between gel and film during exposure	Make sure the X-ray cassette is flat during exposure

PROBLEM	POSSIBLE CAUSES	SOLUTION
Bands curve up near the sides (smiling)	Nonuniform heating of gel during electrophoresis	Pre-run the gel for at least 30 min to preheat it
	Thickness of the gel was not uniform	Clamp directly over spacers during the polymerization
Nonspecific spotting on the film	Radioactive contamination on gel dryer, intensifying screens, gloves, or film cassettes	Check for contamination with a Geiger counter
	Stray radiation near the film during exposure	Expose autoradiogram between glass plates and away from other radiation sources

11F. Computer Analysis

The biological information encoded in a DNA sequence may be determined by searching the sequence for specific sites or comparing the sequence with other DNA sequences. Plodding through the data by eye is extremely tedious, so computers are extensively used for DNA sequence analysis. Learning to analyze DNA sequences on the computer is probably as useful as learning to run the sequencing reactions. Therefore, you should enter your DNA sequence into the computer and try to analyze it as thoroughly as possible. Since your sequence may be relatively short, some of the programs will not be applicable. Therefore, also try analyzing another sequence stored in the DNA database (e.g., the *lac* operon). Different software packages have different features and directions, so the specific instructions will vary depending upon the software available. Like working in the lab, you cannot learn how to use computers without experimenting. An important point to remember is "Do not expect your computer to tell you the truth" (von Heinje, 1987): not all sequence similarity indicates homology and not all sequences that look like promoters function in vivo. Always interpret the output from the computer based on what is known about the molecular genetics of the system.

Try the following DNA sequence analysis programs:

1. Translate the DNA sequence into protein. Determine the potential amino acid sequence of all six possible reading frames.
2. Often codon usage indicates the relative expression of a protein. Determine the codon usage of long reading frames that may encode an actual protein.
3. Predict the secondary structure of the protein (α-helix, β-sheet, turns, hydrophobicity, etc.).
4. Search for consensus regulatory sites (promoters, Rho-independent terminators, ribosome binding sites, CRP binding sites, etc.).
5. Compare with other sequences for sequence similarities, homology, and alignment. Search the GenBank database for similar sequences.

References

Bishop, M., and C. Rawlings. 1987. *Nucleic Acid and Protein Sequence Analysis: A Practical Approach.* IRL Press, Washington, D.C.

Doolittle, R. F. 1986. *Of Urfs and Orfs: A Primer on How to Analyze Derived Amino Acid Sequences.* University Science Books, Mill Valley, CA.

Maloy, S., and S. Olson. 1989. DNAzoom: Educational software for the analysis of DNA and protein sequences. *Academic Computing*, April, pp. 18-50.

von Heijne, G. 1987. *Sequence Analysis in Molecular Biology: Treasure Trove or Trivial Pursuit.* Academic Press, NY.

Roe, B. 1988. Computer programs for molecular biology: An overview of DNA sequencing and protein analysis packages. *BioTechniques* 6: 593-596.

RESULTS
EXPERIMENT 11

1. Include a photocopy of the agarose gel showing your ssDNA yield.
2. Include the autoradiogram of the DNA sequence.
3. List the DNA sequence read from your clone and the M13 control.
4. Compare the sequence of the M13 control with the known sequence.
5. Include any additional computer analysis.

APPENDICES

Appendix 1. Media

E Medium
To make 50x E stock:

10 g	$MgSO_4 7H_2O$
100 g	Citric acid $1H_2O$ (Granular)
655 g	$K_2HPO_4 3H_2O$ (potassium phosphate dibasic)
175 g	$NaNH_4HPO_4 4H_2O$ (sodium ammonium phosphate)

Heat 500 ml dH_2O on stirring block but do not boil.
Add the chemicals one at a time, allowing each to dissolve completely before adding the next.
Bring to 1000 ml with dH_2O.
Cool in refrigerator.
Add about 25 ml chloroform. Store at room temperature.

To make E + glucose plates:
Add 20 ml of 50x E to 500 ml dH_2O. Autoclave.
Add 15 g Bacto-agar to 500 ml dH_2O. Autoclave.
After autoclaving add 10 ml of 20% glucose.
Swirl to mix the solution and pour plates.

To make 1x E medium:
Add 20 ml of 50x E to 980 ml dH_2O. Mix.
Dispense 100 ml in bottles. Autoclave.
For E + glucose medium add 1 ml of 20% glucose per bottle after autoclaving.

Green Plates
For 3000 g Green plate mix:

828	g	Tryptone
105	g	Yeast extract
516	g	NaCl
1151	g	Bacto Agar
6.8 g		Blue Aniline
64.4 g		Alizarin Yellow G,GG

To prepare Green plates from the mix:
Dissolve 29 g Green plate mix in 600 ml dH_2O. Autoclave.
After autoclaving add 400 ml sterile dH_2O and 33.5 ml of 20% glucose before pouring plates.

L-Broth (LB)
Either NB or LB can be used as a rich medium. LB is slightly less expensive and NB is slightly easier to prepare.

10 g	Tryptone
5 g	Yeast extract
5 g	NaCl

Dissolve in 1 liter dH_2O. Autoclave.
To make LB plates add 15 g agar per liter before autoclaving.

MacConkey-Lactose plates

MacConkey medium contains pH indicators that can be used to differentiate Lac⁺ from Lac⁻ colonies: Lac⁺ colonies ferment lactose and the acid turns the pH indicator red colored, but Lac⁻ colonies cannot ferment lactose so they remain white colored. MacConkey plates are less expensive than Xgal plates but Xgal plates are much more sensitive.

    40 g   MacConkey agar base (Difco)
    900 ml dH_2O

Mix thoroughly. Autoclave.
Before pouring plates, add 100 ml of sterile 10% lactose.

NCE

NCE is a minimal medium without citrate. Since *S. typhimurium* can use citrate as a carbon source, this is a good medium for selecting for growth on alternative carbon sources.

To make 50x NCE stock:

    197 g KH_2PO_4 (potassium phosphate monobasic)
    323 g $K_2HPO_4 \, 3H_2O$ (potassium phosphate dibasic)
    175 g $NaNH_4HPO_4 \, 4H_2O$ (sodium ammonium phosphate)

Heat 500 ml dH_2O on stirring block but do not boil.
Add the chemicals one at a time, allowing each to dissolve completely before
    adding the next.
Bring to 1000 ml with dH_2O.
Cool in refrigerator.
Add about 25 ml chloroform. Store at room temperature.

To make NCE plates:

Add 20 ml of 50x NCE to 500 ml dH_2O. Autoclave.
Add 15 g agar to 500 ml dH_2O. Autoclave.
After autoclaving add 1 ml of 1 M $MgSO_4$ and a sterile carbon source to 0.2%.
    Swirl to mix the solution and pour plates.

Nutrient Broth (NB)

Either NB or LB can be used as a rich medium. LB is slightly less expensive and NB is slightly easier to prepare.

    8 g   Nutrient Broth (Difco)
    5 g   NaCl

Dissolve in 1 liter dH_2O. Autoclave.
To make NB plates add 15 g agar per liter before autoclaving.

Plasmid broth

Plasmid broth is a very rich, buffered medium that allows cells to grow to a high density and gives an excellent yield of plasmid DNA.

    12 g   Tryptone
    24 g   Yeast extract
    5 ml   Glycerol
    900 ml dH_2O

After autoclaving add 100 ml of 1 M Potassium phosphate buffer pH 7.6 (see
    below).
Add an appropriate antibiotic (e.g. 20 mg chloramphenicol or 25 mg sodium
    ampicillin) just before use.

1 M Potassium phosphate buffer pH 7.6

    69 g KH_2PO_4 (potassium phosphate monobasic)

    300 ml dH_2O

Adjust to pH 7.6 with KOH. Bring to 500 ml with dH_2O. Dispense into 100 ml aliquots and autoclave.

TS Top Agar

    10 g   Tryptone

    8 g   NaCl

    7 g   Agar

    1000 ml dH_2O

Dissolve in a microwave, dispense in bottles, and autoclave.

Appendix 2. Solutions

Acrylamide:Bis-acrylamide Stock (30: 0.8). For SDS-Polyacrylamide gels.
CAUTION - ACRYLAMIDE IS TOXIC. AVOID CONTACT WITH SKIN.

      30.0 g   Acrylamide
       0.8 g   Bis-acrylamide

Dissolve in about 75 ml ddH$_2$0.
Bring to 100 ml with ddH$_2$0.
Filter through Whatman #1 filter paper.
Store in a brown bottle at 4°C.

Acrylamide/bis-acrylamide stock (38: 2). For DNA gels.
CAUTION - ACRYLAMIDE IS TOXIC. AVOID CONTACT WITH SKIN.

      76 g   Ultra-pure acrylamide
       4 g   Bis-acrylamide

Dissolve in about 150 ml ddH$_2$0.
Bring to 200 ml with ddH$_2$0.
Filter through Whatman #1 filter paper.
Store in a brown bottle at 4°C.

NOTE: Acrylamide slowly breaks down into acrylic acid which affects the polymerization and electrophoretic properties of polyacrylamide gels. Acrylamide solutions can be deionized to remove any acrylic acid as follows:
Mix 100 ml acrylamide / bis-acrylamide solution with 10 g mixed bed ion-exchange resin (Bio-rad AG501-XB, 20-50 mesh).
Stir for 30 min at room temperature then store at 4°C.
Before use, filter through Whatman #1 paper to remove the resin.

7.5 M Ammonium acetate

Dissolve 57 g ammonium acetate in about 80 ml ddH$_2$0.
Bring to 100 ml with ddH$_2$0.
Sterilize by passing through a 0.2 μm membrane filter.
Store at room temperature.

Ammonium persulfate (100 mg/ml)

Dissolve 1 g ammonium persulfate in 8 ml ddH$_2$0.
Bring to 10 ml with ddH$_2$O.
Aliquot and store at -20°C.

100 mM ATP

      60 mg   ATP
      800 μl   ddH$_2$0

Neutralize to pH 7 with 0.1 M NaOH. (NTPs undergo acid catalyzed hydrolysis if they are not neutralized.)
Bring to 1 ml with ddH$_2$0.
Store frozen in 0.1 ml aliquots at -20°C.

Blue II

Use a 1:10 dilution as a loading buffer for electrophoresis of DNA. Glycerol increases the density of the sample allowing it to be loaded into submerged wells. Xylene cyanol and bromphenol blue are useful tracking dyes.

25 ml	100% Glycerol
5 ml	0.5 M EDTA
20 ml	ddH_2O
50 mg	Xylene cyanol
75 mg	Bromphenol blue

10 mg/ml BSA (Bovine serum albumin)

BSA is included in many reactions because it stabilizes many enzymes and it binds nonspecifically to many surfaces, decreasing nonspecific binding by critical components of the reaction. Pentax fraction V is a pure grade of BSA that usually lacks nucleases and proteases. Certified nuclease-free BSA can also be purchased but it is much more expensive.

Dissolve 10 mg BSA (Pentax Fraction V) in 1.0 ml sterile ddH_2O.

Freeze 0.1 ml aliquots at -20°C.

Dilute 1/100 in ddH_2O and check the A_{280}. The A_{280} of the diluted BSA in a 1 cm cuvette should equal 0.066.

1 M CaCl$_2$

Dissolve 14.7 g CaCl$_2$ 2H$_2$0 in about 80 ml ddH$_2$0.

Bring to 100 ml with ddH$_2$0 and autoclave.

50 mM CaCl$_2$

1.5 ml	sterile 1 M CaCl$_2$
28.5 ml	sterile ddH_2O

Chloroform:isoamyl alcohol (24:1)

Add 120 ml chloroform and 5 ml isoamyl alcohol to a brown bottle. Mix and store at room temperature.

D-cycloserine (**Prepare fresh**)

50 mg	D-cycloserine
1 ml	sterile ddH_2O

Defined amino acids minus methionine (DAA)

Mix 5 ml of each of the following amino acid stock solutions.

alanine	glycine	leucine	threonine
arginine	glutamate	lysine	tryptophan
aspartate	histidine	proline	tyrosine
cysteine	isoleucine	serine	valine

Concentrations of the amino acid stock solutions are shown in Appendix 4.

Add 40 ml DAA to 210 ml E medium to obtain the final concentration of amino acids shown in Appendix 4.

100x Denhardt's solution

0.5 g	Ficoll (M$_r$ 400,000)
0.5 g	Polyvinyl pyrrolidone (M$_r$ 360,000)
0.5 g	BSA (Bovine serum albumin, Pentax fraction V)

Add to 25 ml of 2x SSC. Stir to dissolve.

Aliquot and store frozen at -20°C.

dH$_2$O *and* ddH$_2$O
The purity of the water used can have a significant effect on many experiments. All media and most solutions can be made with deionized or distilled water (dH$_2$O). When ddH$_2$O is indicated, highly purified water (e.g., distilled-deionized water) should be used.

DNase I stock solution **(1 mg/ml)**

5.0 mg	DNase I (RNase-free, 2000-2500 units/mg)
2.5 ml	Sterile glycerol
0.5 ml	1 M TrisHCl pH 7.4
2.0 ml	Sterile ddH$_2$0

Dissolve the DNase.
Store in 100 µl aliquots at -20°C.
Thaw on ice for about 1 hr making the working solution.

DNase I diluted working solution

10 ml	ddH$_2$0
100 µl	1 M TrisHCl pH 7.4
50 µl	1 M MgCl$_2$
1 µl	1 mg/ml DNase I

Dilute immediately prior to use.

dNTP solution for nick translation

10 µl	200 µM dCTP
10 µl	200 µM dGTP
10 µl	200 µM dTTP

Depurination solution (0.2 M HCl) for Southern blots

10 ml	6 N HCl
290 ml	ddH$_2$O

Denaturation solution (0.5 M NaOH, 1.5 M NaCl) for Southern blots

20 g	NaOH
88 g	NaCl

Dissolve in about 800 ml dH$_2$O.
Bring to 1000 ml with dH$_2$O.

Dialysis tubing
Always wear gloves when handling dialysis tubing.
Cut the dialysis tubing into 10-20 cm lengths.
Boil for 10 min in a large beaker of 2% sodium bicarbonate and 1 mM EDTA.
Rinse the tubing thoroughly with dH$_2$O.
Boil for 10 min in a large beaker of 1 mM EDTA.
Allow to cool. Store in the 1 mM EDTA solution at 4°C.
Before use, thoroughly rinse the tubing inside and out with dH$_2$O.

1 M DTT **(Dithiothreitol)**
Most sulfhydryl groups in bacterial enzymes exist in a reduced form. When exposed to oxygen, the sulfhydryl groups may be oxidized to disulfides, inactivating the enzyme. DTT [HS-CH$_2$(CHOH)$_2$CH$_2$-SH] reduces disulfide bonds to thiols.
Dissolve 309 mg DTT in 2 ml ddH$_2$O.
Freeze 0.1 ml aliquots at -20°C.

0.5 M EDTA (Ethylenediamine tetraacetic acid)
EDTA chelates divalent cations.

Add 18.6 g Na$_2$EDTA 2H$_2$O to about 70 ml ddH$_2$0 on a magnetic stirrer.
Slowly adjust to pH 8.0 with NaOH (about 5 ml of 10 N NaOH). EDTA will not go into soution until the pH is near 8.
Bring to 100 ml with ddH$_2$0. Autoclave.

1 M EGTA
Ethylene glycol-bis(ß-aminoether)N,N,N',N'-tetra acetic acid (EGTA) is a chelating agent that specifically complexes with Ca^{++} but not Mg^{++} (Schmid, R., and C. Reilly. 1957. *Anal. Chem.* 29: 264-268).

Add 250 g EGTA to about 400 ml dH$_2$0 on a magnetic stirrer.
Add 53 g NaOH.
Continue stirring until completely dissolved.
Bring to pH 7 with HCl.
Bring to 660 ml with dH$_2$0.
Dispense in bottles. Autoclave.
Use 10 ml per liter medium or spread 0.2 ml per plate.

Electroelution buffer
(200 mM NaCl, 50 mM Tris, 1 mM EDTA, 0.1% SDS)

400	µl	5 M NaCl
500	µl	1 M TrisHCl (pH 8)
20	µl	0.5 M EDTA
100	µl	10% SDS
10	ml	ddH$_2$O

10 mg/ml Ethidium bromide
Ethidium bromide intercalates between the stacked bases of DNA and RNA. When excited by UV light between 254 nm (short wave) and 366 nm (long wave) it emits fluorescent light at 590 nm. Due to the decreased rotation possible when it is intercalated, the DNA-ethidium bromide complex produces about 50 times more fluorescence than free ethidium bromide. Ethidium bromide can be used to detect both double stranded and single stranded nucleic acids, but the sensitivity is much greater for double stranded DNA. When ethidium bromide stained gels are photographed on a UV transilluminator a UV filter is required to screen out the background UV from the transilluminator. Ethidium bromide is a known carcinogen: based on the Ames test, 90 µg of ethidium bromide is about as carcinogenic as one cigarette.

CAUTION-ETHIDIUM BROMIDE IS A CARCINOGEN. WEAR GLOVES.
0.2 g ethidium bromide
Dissolve in 20 ml ddH$_2$0 on a magnetic stirrer.
Store in a dark bottle at 4°C.

Freezer vials
DMSO (Dimethylsulfoxide) decreases damage to cells due to ice crystal formation during freezing and thawing. When stored frozen in DMSO at -70°C, most bacteria remain viable for many years.

Add 0.2 ml DMSO to each 2 ml vial. **Wear gloves.**
Loosely cap and autoclave for 15 min.
Tighten caps and store at room temperature.
To use, fill vial with about 1.5 ml of a fresh culture, invert to mix, label, and place in -70°C freezer.
To revive a frozen culture, scrape a small chip of ice from the vial with a sterile stick and streak on an NB plate. Do NOT allow the culture to thaw.

Formamide (deionized)

Formamide slowly breaks down into formic acid and ammonia which can cause degradation of nucleic acids. The formamide must be deionized to remove the breakdown products: formamide is not charged so it does not bind to the resin, but the breakdown products are charged so they bind tightly to the ion-exchange resin.

> Mix 100 ml formamide with 10 g mixed bed ion-exchange resin (Bio-rad AG501-XB, 20-50 mesh).
> Stir for 30 min at room temperature then store at 4°C.
> Before use, filter through Whatman #1 paper to remove the resin.

20% Glucose

> Heat 250 ml dH_2O on a stirring block but do not boil.
> Slowly add 100 g D-glucose to the hot dH_2O.
> After the glucose is completely dissolved, bring to 500 ml with dH_2O.
> Dispense into bottles and autoclave.

Hybridization solution for Southern blots

> 4.0 ml 20x SSC
> 1.0 ml 100x Denhardt's solution
> 1.0 ml 10% SDS
> 0.4 ml 5 mg/ml salmon sperm DNA
> 10.0 ml Formamide (final concentration = 50%)
> 3.6 ml ddH_2O

Mix and filter through a 0.2 μm membrane filter.

Hydroxylamine (prepare fresh)

> 0.175 g Hydroxylamine (NH_2OH)
> 0.28 ml 4 M NaOH

Bring to 2.5 ml with sterile dH_2O.

IPTG (**isopropyl-ß-D-thiogalactoside**) is a gratuitous inducer of the *lac* operon.

> Add 5 mg IPTG to 1 liter medium after autoclaving or if only a few plates are needed dissolve 5 mg IPTG in 1 ml sterile dH_2O and spread 50 μl per plate.
> The stock solution can be stored frozen at -20°C.

10% Lactose

> Dissolve 10 g lactose in 80 ml dH_2O.
> Bring to 100 ml with dH_2O. Autoclave.

LBSE

> 100 ml LB
> 0.2 ml 0.5 M EDTA
> 5.85 g NaCl

Mix. Autoclave. Store at 4°C.

10x Ligase buffer for T4 DNA ligase

> 0.5 ml 1 M TrisHCl pH 7.6
> 0.1 ml 1 M $MgCl_2$
> 0.1 ml 100 mM DTT
> 0.1 ml 10 mg/ml BSA (Bovine serum albumin, Pentax fraction V)
> 0.1 ml 100 mM ATP
> 0.1 ml sterile ddH_2O

Store frozen in small aliquots at -20°C.

Lysis solution for plasmid purifications
      4.0 ml ddH_2O
      0.25 ml 1 M Tris HCl pH 8
      0.20 ml 0.5 M Na_2EDTA
      0.50 ml 20% glucose

1 M $MgCl_2$
      Dissolve 20.3 g $MgCl_2$ $6H_2O$ in 80 ml ddH_2O.
      Bring to 100 ml with ddH_2O and autoclave.

1 M $MgSO_4$
      Dissolve 24.6 g $MgSO_4$ $7H_2O$ in 100 ml dH_2O.
      Autoclave.

Neutralization solution (1 M TrisHCl, 1.5 M NaCl) for Southern blots
      13.4 g   Tris base
    140.4 g   TrisHCl
      88   g   NaCl
    Dissolve in 800 ml dH_2O.
    Bring to 1000 ml with dH_2O.

10x Nick translation buffer
      50 µl   1 M TrisHCl pH 7.4
      10 µl   1 M $MgCl_2$
       1 µl   1 M DTT
       1 µl   100 mg/ml BSA (Bovine serum albumin, Pentax Fraction V)
      38 µl   Sterile ddH_2O

ONPG (4 mg/ml) (Sufficient for 100 ß-galactosidase assays)
      80 mg   o-nitrophenyl-ß-D-galactoside (o-nitrophenyl-ß-D-galacto-
                    pyranoside)
      20 ml dH_2O

P22 Broth (store at 4°C)
      100 ml   NB
        2 ml   50x E medium
        1 ml   20% glucose
      0.1 ml   P22 HT *int* phage (approximately 5×10^{10} pfu/ml)

PEG/NaCl
      20.0 g Polyethylene glycol (8000 MW)
      14.6 g NaCl
    Dissolve in about 60 ml ddH_2O.
    Bring to 100 ml with ddH_2O.
    Store at room temperature.

Phenol (TE saturated)

CAUTION - PHENOL CAN CAUSE SEVERE BURNS. WEAR SAFETY GLASSES AND GLOVES. DO NOT MOUTH-PIPET. If phenol contacts your skin, immediately and thoroughly rinse with water and notify the instructor. Do NOT rinse with alcohol.

Remove redistilled phenol from the freezer and warm to room temperature. Warm the phenol in a 43-60°C water bath to melt it. The phenol should be colorless. Do not use if the phenol is colored.

Add 8-Hydroxyquinoline to 0.1% final concentration (yellow colored).

Add an equal volume of 1 M Tris HCl pH 8 to saturate the phenol, mix, and remove the aqueous phase. Check the pH of the aqueous phase. If necessary, continue extracting the phenol until the pH of the aqueous phase is greater than 7.

Aliquot 1 ml of the phenol solution into microfuge tubes. Add 0.3 ml TE to each tube. Mix by inverting.

Store the tubes in a brown bottle at -20°C until use.

Phosphate-EDTA buffer (0.5 M KPO_4 pH 6, 5 mM EDTA) for hydroxylamine mutagenesis

6.81 g	KH_2PO_4
70 ml	dH_2O

Dissolve on a stirrer.
Bring to pH 6 with 1 N KOH.
Bring to 99 ml with dH_2O.
Add 1 ml of 0.5 M EDTA.
Autoclave.

Potassium acetate pH 4.8 (3 M potassium, 5 M acetate)

29.4 g	Potassium acetate
60.0 ml	ddH_2O

Dissolve then add:

11.5 ml	Glacial acetic acid
28.5 ml	ddH_2O

20 mg/ml Proteinase K

Proteinase K is a nonspecific serine protease that has strong proteolytic activity even in the presence of SDS. It rapidly inactivates endogenous nucleases in the lysed cells.

Dissolve 10 mg proteinase K in 500 µl ddH_2O.
Store 20 µl aliquots at -20°C.

1 mg/ml RNase (DNase free)

1 mg	Ribonuclease A
10 µl	1 M TrisHCl pH 7.4
3 µl	5 M NaCl
987 µl	ddH_2O

Place in boiling water bath for 15 min. Remove and allow to slowly cool to room temperature. Dispense in aliquots and store at -20°C.

Dilute to 1 µg/ml before use. Test each batch of RNase to make sure there is no contaminating DNase as follows:

Mix 8 µl sterile ddH_2O + 1 µl Lambda *Hin* dIII standards + 2 µl RNase, incubate 1 hr at 37°C, add 1 µl Blue II and run on a 0.8% agarose gel alongside untreated Lambda *Hin* dIII standards.

Salmon sperm DNA (5 mg/ml)
Salmon sperm DNA is often used to saturate any nonspecific DNA binding sites during hybridization experiments. To be effective, it must be sheared into shorter fragments.

> Dissolve 50 mg DNA in 10 ml sterile ddH$_2$0 in a sterile test tube (Type-III DNA sodium salt from salmon testes, Sigma Chemical Co).
> Vortex vigorously to dissolve.
> Pass several times through an 18 gauge hypodermic needle to shear.
> Phenol extract 2x with an equal volume of phenol:chloroform:isoamyl alcohol (25:24:1).
> Denature in a boiling water bath for 15 min.
> Quick cool on ice. Store 0.5 ml aliquots at -20°C.
> Determine the fragment size on a 1.2% agarose gel (Appendix 7A). An average fragment size of about 700 bp is ideal for hybridizations.

SDS-NaOH (Prepare fresh. Do not chill)

9.0 ml	ddH$_2$0
0.2 ml	10 N NaOH
1.0 ml	10% SDS

SDS-Sample Buffer (4% SDS, 62.5 mM Tris pH 6.8, 10% Glycerol, 8 M Urea, 0.010% Bromphenol Blue) SDS and urea denature proteins, glycerol increases the density of the solution so it is easier to load onto a gel, and bromphenol blue is a tracking dye.

1.0 g	SDS
2.5 ml	0.625 M Tris pH 6.8
2.5 ml	Glycerol
2.0 mg	Bromphenol Blue
12.0 g	Urea
0.5 g	DTT

Bring to 25 ml with ddH$_2$O. Aliquot into microfuge tubes and store at -20°C.

Sephadex G-50 columns
Sephadex is used for gel filtration columns that separate molecules based upon size and shape. Large molecules elute from the column quickly while small molecules become entrapped in the Sephadex beads and elute much slower.

> Hydrate the resin by adding 10 g Sephadex G-50 (medium) to 100 ml TE and autoclaving for 15 min.
> Allow to cool at room temperature. Swirl gently, allow to settle a few min, then pour off the excess TE and any fine Sephadex particles that don't sediment. Add fresh TE and store at 4°C until use.
> Use small (6 ml) fritted bottom dispo-columns or Pasteur pipets plugged with a small amount of glass wool. Swirl the flask of Sephadex to form a uniform suspension and fill the column with the suspension.
> Allow the Sephadex to settle then add more until the column contains about 4 ml of packed Sephadex. Packing the column takes about 5-10 min. The Sephadex columns can be stored at 4°C.

3 M Sodium acetate

> Dissolve 40.8 g NaAcetate 3H$_2$O in about 80 ml ddH$_2$O.
> Adjust to pH 5.2 with glacial acetic acid.
> Bring to 100 ml with ddH$_2$O. Autoclave.

1 M Sodium carbonate (Sufficient for 100 ß-galactosidase assays)

5.3 g	Na$_2$CO$_3$
50 ml	dH$_2$O

0.85% Sodium chloride
> Dissolve 8.5 g NaCl in 1 liter dH_2O.
> Dispense in bottles. Autoclave.

5 M Sodium chloride
> Dissolve 29.2 g NaCl in 80 ml ddH_2O.
> Bring to 100 ml with ddH_2O. Autoclave.

10% Sodium dodecyl sulfate (SDS)
> Dissolve 10 g SDS in 90 ml ddH_2O. (May need to be warmed to dissolve).
> Adjust to pH 7.2 with a few drops of 6 N HCl.
> Bring to 100 ml with ddH_2O.

10 N Sodium hydroxide
> Slowly dissolve 40 g NaOH pellets in about 60 ml ddH_2O on a magnetic stirrer.
> **(CAUTION—This is an exothermic reaction and the solution gets very hot!)**
> Bring to 100 ml with ddH_2O.

20x SSC (3 M NaCl, 0.3 M sodium citrate)
> 175.3 g   NaCl
> 88.2 g Na_3Citrate $2H_2O$
> Dissolve in 800 ml ddH_2O.
> Adjust to pH 7.0 with HCl.
> Bring to 1000 ml with ddH_2O.
> Add 20 ml of 20x SSC to 180 ml ddH_2O for 2x SSC working solution.

T2 buffer. This is a good buffer for storing dilute phage stocks. The pH, ionic strength, cation concentration, and gelatin help stabilize phage structural proteins.
> 2.5 g K_2SO_4
> 2.0 g   NaCl
> 0.75 g KH_2PO_4 (potassium phosphate monobasic)
> 1.5 g Na_2HPO_4 (sodium phosphate dibasic)
> Dissolve in 500 ml dH_2O.
> Aliquot 100 ml into bottles. Autoclave.
> After autoclaving, cool to room temperature, then add 0.33 ml of each of the following sterile solutions per 100 ml:
> 0.1%   gelatin
> 0.1 M $MgSO_4$
> 0.01 M $CaCl_2$

10x TBE. TBE is a good buffer for electrophoresis of DNA. The stock solution should be diluted to 1x before use.
> 16 g   Tris base
> 27.2 g   Boric acid
> 9.3 g Na_2EDTA
> Dissolve in about 600 ml ddH_2O on a magnetic stirrer.
> Bring to 1 liter with ddH_2O. Filter through Whatman #1 filter paper.
> Store at room temperature. Discard if a solid precipitate develops.

10% TCA (Trichloroacetic acid)
> Dissolve 10 g Trichloroacetic acid in about 80 ml ddH_2O.
> Bring to 100 ml with ddH_2O.

TE (10 mM Tris, 1 mM EDTA pH 8)

1.0 ml	sterile 1 M TrisHCl pH 8.0
0.2 ml	sterile 0.5 M EDTA
99.0 ml	sterile ddH$_2$0

Mix the solutions in a sterile bottle and store at room temperature.

Toothpick lysis solution

6 ml	ddH$_2$O
3 ml	10% SDS
1 ml	1 M Tris HCl pH 8
120 μl	10 N NaOH

1 M Tris buffers

The pH of Tris buffers changes significantly at different temperatures, so the temperature of the reaction must be taken into consideration when preparing the buffer. (The pH of Tris buffers at room temperature is usually indicated on the solutions.) Always double-check the final pH after preparing buffer solutions. Sterilize the solutions by autoclaving.

Final pH			Tris Base	Tris HC1
5°C	25°C	37°C	(g / l)	(g / l)
7.76	7.2	6.91	13.4	140.4
7.97	7.4	7.12	19.4	132.2
8.18	7.6	7.30	27.8	121.2
8.37	7.8	7.52	39.4	106.4
8.58	8.0	7.71	53.0	88.8
8.78	8.2	7.91	66.8	70.8

Reference: Sigma Chemical Company Technical Bulletin #106B

Xgal (5-bromo-4-chloro-3-indoyl-ß-D-galactoside) is a specific indicator for ß-galactosidase activity: Lac$^+$ cells form blue colonies and Lac$^-$ cells form white colonies.

20 mg	Xgal
1 ml	N,N-dimethylformamide.

Vortex to dissolve.

Add 1 ml per liter of medium after autoclaving or if only a few plates are needed spread 50 μl of the Xgal solution per plate and allow to soak in thoroughly.

Z-buffer stock solution for ß-galactosidase assays

4.27 g	Na$_2$HPO$_4$
2.75 g	NaH$_2$PO$_4$ H$_2$O
0.375 g	KCl
0.125 g	MgSO$_4$ 7H$_2$O

Adjust to pH 7.0.

Bring to 500 ml with dH$_2$O.

Do not autoclave. Store at 4°C.

For complete Z-buffer — Prior to daily use mix:

50 ml	Z-buffer
0.14 ml	ß-mercaptoethanol

Appendix 3. Antibiotics

A. ANTIBIOTIC
CONCENTRATIONS

| Antibiotic | Final concentration[1] | | Stock solution[2] |
	Rich media	Minimal media	
Na Ampicillin (Amp)[3]	30 µg/ml	15 µg/ml	9 mg/ml
Chloramphenicol (Cml)	20 µg/ml	5 µg/ml	6 mg/ml
Kanamycin SO_4 (Kan)[4]	50 µg/ml	125 µg/ml	15 mg/ml
Tetracycline HCl (Tet)	20 µg/ml	10 µg/ml	6 mg/ml
Streptomycin SO_4 (Str)[5]	------	2 mg/ml	——

[1]The solid form of the antibiotics can be added directly to sterilized media that has been cooled to approximately 55°C. If kept at 4°C tetracycline, chloramphenicol, and streptomycin plates are usually good for several months, but kanamycin plates and ampicillin plates may only last for several weeks. A sensitive and resistant control should always be included each time antibiotic plates are used.

[2]For liquid media or for a few plates, a stock solution of the antibiotics can be prepared and stored at -20°C. All the antibiotic stock solutions can be prepared in sterile dH_2O except chloramphenicol which can be dissolved in dimethylformamide. The stock solution is 300x the concentration of antibiotic required in rich medium. An average petri dish contains about 30 ml medium, so spread 0.1 ml on rich plates.

[3]High level expression of ß-lactamase can destroy ampicillin in the medium surrounding the Amp[r] colonies, allowing Amp[s] satellite colonies to grow. Satellite colonies can be decreased by using 2x ampicillin when selecting for plasmids.

[4]To select for a Kan[r] gene on a multicopy plasmid use LB + 10 mM Tris HCl pH 7.4 + 250 µg/ml Kanamycin SO_4 or Neomycin SO_4. High level Kan[r] cannot be selected if the pH of the medium is less than 7.2 (P. Berget, R. Maurer, and G. Weinstock, personal communication).

[5]For streptomycin plates add 1% Nutrient broth to the minimal media (i.e., 1 ml NB/ 100 ml medium).

B. ANTIBIOTIC SENSITIVITY AND RESISTANCE

Ampicillin inhibits crosslinking of peptidoglycan chains in the cell wall of eubacteria. Cells growing in the presence of ampicillin synthesize weak cell walls, causing them to burst due to the high internal osmotic pressure. Amp^r encoded by Mu derivatives and pBR plasmids is due to a periplasmic ß-lactamase that breaks the ß-lactam ring of ampicillin.

Chloramphenicol inhibits protein synthesis by binding to the 50s ribosomal subunit and blocking the peptidyltransferase reaction. Cam^r encoded by pBR328 is due to a cytoplasmic chloramphenicol acyltransferase which inactivates chloramphenicol by covalently acetylating it.

Kanamycin inhibits protein synthesis by binding to the 30s ribosomal subunit and preventing translocation. Kan^r is usually due to a cytoplasmic aminoglycoside phosphotransferase that inactivates kanamycin by covalently phosphorylating it. Kan^r requires phenotypic expression. Neomycin is a structural analog of Kanamycin that functions by the same mechanism and is inactivated by the same mechanism.

Tetracycline inhibits protein synthesis by preventing aminoacyl tRNA from binding to ribosomes. Tet^r encoded by Tn*10* and pBR plasmids is due to a membrane protein that actively transports tetracycline out of the cell. When Tn*10* is present in multiple copies, cells are less resistant to Tetracycline than when only one copy of Tn*10* is present.

Streptomycin inhibits protein synthesis by binding to the S12 protein of the 30s ribosomal subunit and inhibiting translation. A high level of Str^r can result from chromosomal mutations in the gene for the S12 protein (*rpsL*) which prevent streptomycin from binding to the ribosome. Since only mutant ribosomes are Str^r, resistance to streptomycin is recessive to streptomycin sensitivity. Str^r requires phenotypic expression.

Reference

Foster, T. 1983. Plasmid determined resistance to antimicrobial drugs and toxic metal ions in bacteria. *Microbiol. Rev.* 47:361-409.

Appendix 4. Concentrations of Nutritional Supplements[1]

Nutrient	Plate Concn.	% Stock Solution[2]	Sterilize[3]	Comments
Adenosine	0.5	2.67	Autoclave	
Alanine	0.47	0.84	Autoclave	
p-Aminobenzoic acid (PABA)	0.02	0.06	Filter	Store in dark at 4°C
∂-Aminolevulinic acid HCl (∂-ALA)	0.3	1.00	Filter	
Arginine HCl	0.6	2.5	Autoclave	
Asparagine H_2O	0.32	0.96	Filter	
Aspartate, Na H_2O	0.3	1.04	Filter	
Biotin		0.0024	Filter	Store at 4°C
Cysteine HCl H_2O	0.3	1.05	Filter	Store in dark
Diaminopimelic acid (DAP)	0.1	0.38	Autoclave	
Dihydroxybenzoic acid (DHBA)	0.02	0.06	Filter	Store at 4°C
Glutamate, Na	5.0	18.7	Filter	
Glutamine	5.0	14.6	Filter	In 1 N HCl
Glycine	0.13	0.2	Autoclave	
Guanosine	0.3	1.7	Autoclave	In 1 N HCl
Histidine HCl H_2O	0.1	0.42	Autoclave	
Isoleucine	0.3	0.79	Autoclave	
Leucine	0.3	0.79	Autoclave	
Lysine	0.3	1.1	Autoclave	
Methionine	0.3	0.9	Autoclave	
Nicotinic acid	0.1	0.25	Autoclave	
Pantothenate, Ca	0.1	0.57	Filter	Store at 4°C
Phenylalanine	0.3	0.99	Autoclave	In 0.01 N HCl
Proline	2.0	4.6	Autoclave	
Pyridoxine HCl	0.1	0.41	Autoclave	
Serine	4.0	8.4	Autoclave	
Thiamine H_2O (B1)	0.05	0.355	Filter	
Threonine	0.3	0.71	Autoclave	
Thymine	0.32	0.81	Autoclave	
Tryptophan	0.1	0.41	Filter	Store in dark
Tyrosine	0.1	0.36	Filter	In 1 N NaOH in dark
Uracil	0.1	0.224	Autoclave	
Valine	0.3	0.7	Autoclave	

[1] Modified from Davis, R., D. Botstein, and J. Roth. 1980. *Advanced Bacterial Genetics*, p. 206. Cold Spring Harbor Laboratory, NY.

[2] The stock solutions are 200x. Use 5 ml of the stock solution per liter of medium or spread 0.1 ml per plate.

[3] When "Filter" is indicated, the solutions should be sterilized by filtering through a 0.2 μm millipore filter and stored in sterile bottles. Light sensitive solutions should be stored in sterile dark brown or foil wrapped bottles.

Appendix 5. Basic Molecular Biology Techniques

**A. PHENOL
    EXTRACTION**

Phenol extraction is used to remove proteins from aqueous DNA samples. Extraction with phenol:chloroform:isoamyl alcohol (25:24:1) is even better at removing proteins than phenol alone. Phenol and chloroform denature proteins. The denatured proteins partition into the organic phase or remain at the interphase but the DNA remains in the aqueous phase. Isoamyl alcohol improves the separation of the aqueous and organic phases.

Oxidation of phenol forms quinones (BRL Focus 7:2). Quinones can be further oxidized to yellow and pink colored diacids and phenoxide radicals. These oxidation products interact with $1°$ amines and can cause breakdown or crosslinking of DNA. Therefore, when preparing phenol solutions always check the color and do not use the solution if it has a yellow or pink color. The oxidation products can be removed by distilling the phenol but this is a potentially hazardous and time-consuming job. Several companies sell distilled, recrystallized, "ultra-pure" phenol which seems to work just as well. Pure phenol melts at $43°C$.

When phenol is saturated with water, the aqueous phase often has a low pH due to small amounts of contaminating diacids. Therefore, phenol is usually extracted with buffer before use. The antioxidant 8–hydroxyquinoline is often added to the buffer saturated phenol to decrease the breakdown of phenol. (Hydroxyquinoline is yellow colored which helps identify the phenol phase. This makes it difficult to see the accumulation of oxidation products but if high quality phenol is used initially this does not seem to be a problem.) Buffer saturated phenol is only good for about a month if stored at $4°C$ but small aliquots stored at $-20°C$ seem to be stable for many months. If the phenol is not saturated with an aqueous solution, the aqueous phase of the DNA sample may completely mix with the organic phase. If this happens, addition of a small amount of TE or chloroform:isoamyl alcohol (24:1) usually resolves the two phases. If the salt concentration of the sample is too high the aqueous and organic phases may be inverted but the phenol phase can be identified by its yellow color.

Proteins can be extracted from small aqueous nucleic acid samples in a microfuge tube by the following procedure.

Phenol Extractions

CAUTION - PHENOL CAN CAUSE SEVERE BURNS. WEAR GLOVES.

1. Mix an equal volume of TE saturated phenol with chloroform:isoamyl alcohol (24:1) just before use.
2. Add an equal volume of the phenol:chloroform:isoamyl alcohol (25:24:1) to the DNA sample.
3. Mix thoroughly to form an emulsion.
4. Spin for 3-5 min in a microfuge to separate the phases.
5. Carefully remove the upper aqueous phase, avoiding the interphase or phenol phase. Discard the used phenol in a glass bottle. A good rule of thumb is if you can see any white material at the interphase do another extraction.
6. If any residual phenol remains in the aqueous phase, it can severely inhibit enzymes used in later steps. Ether extraction of the aqueous phase can be done to remove any traces of phenol (see Maniatis et al., 1983) but ether is potentially hazardous and this step is usually not necessary.

Reagents

Chloroform:isoamyl alcohol (24:1)
>   Add 120 ml chloroform and 5 ml isoamyl alcohol to a brown bottle.
>   Mix and store at room temperature.

Phenol (TE saturated)
WEAR SAFETY GLASSES AND GLOVES. DO NOT MOUTH-PIPET. If phenol contacts your skin, immediately and thoroughly rinse with water and notify the instructor. Do NOT rinse with alcohol.
>   Remove redistilled phenol from freezer and allow to warm to room temperature. Warm the phenol in a 43-60°C water bath to melt it. The phenol should be colorless. Do not use if the phenol is colored.
>   Add 8-Hydroxyquinoline to 0.1% final concentration (yellow colored).
>   Add an equal volume of 1 M Tris HCl pH 8 to saturate the phenol, mix, and remove the aqueous phase. Check the pH of the aqueous phase. If necessary, continue extracting the phenol until the pH of the aqueous phase is greater than 7.
>   Aliquot 1 ml of the phenol solution into microfuge tubes. Add 0.3 ml TE to each tube. Mix by inverting. Store the tubes in a brown bottle at -20°C until use.

B. ETHANOL PRECIPITATION

In an aqueous solution with 0.1-0.5 M monovalent cations and 70% ethanol, DNA and RNA precipitate. Ammonium acetate, sodium acetate, and NaCl are commonly used counterions for ethanol precipitation. The salts are usually added to aqueous solutions at the concentrations shown in the following table.

Salt	Stock soln.	Dilution
Ammonium acetate	7.5 M	1/2
Sodium acetate	3.0 M	1/10
Sodium chloride	5.0 M	1/50

Most protocols for ethanol precipitation require cooling the solution to –20°C or –70°C. However, several recent studies have shown that even low concentrations of DNA (<5 ng/ml) are efficiently precipitated at 4°C (BRL Focus 9: 3-5; BRL Focus 7:1-2). These studies also indicate that centrifugation of the precipitated DNA at room temperature may be more efficient than centrifugation at 4°C.

In addition to nucleic acids, salts are also trapped in the pellet. Washing the nucleic acid pellet with 70% ethanol helps remove some of the precipitated salts that may interfere with later reactions. Ammonium acetate is very soluble in ethanol and is effectively removed by a 70% ethanol wash. In contrast sodium acetate and NaCl are less soluble in ethanol so they are removed less efficiently by a 70% ethanol wash. In addition, DNA precipitated in ammonium acetate seems to have fewer contaminants that inhibit restriction enzymes (BRL Focus 4: 12). Therefore, usually ammonium acetate is the best salt for ethanol precipitation of DNA. (However, polynucleotide kinase is strongly inhibited by ammonium and so ammonium acetate precipitation should not be used if the DNA is to be treated with polynucleotide kinase.)

Nucleic acids can also be precipitated by adding an equal volume of isopropanol to aqueous solutions instead of ethanol. However, isopropanol causes much more salt precipitation than ethanol, especially if the solution is chilled. In addition, iso-propanol is less volatile than ethanol so it is much more difficult to remove residual isopropanol after precipitation. Therefore, isopropanol is usually used only if it is necessary to keep the volume of the solution to a minimum.

DNA can be precipitated from small volumes of aqueous solution in a microfuge tube as follows.

Ethanol Precipitation

1. Add 0.5 volume of 7.5 M ammonium acetate.
2. Add 2.5 times the total volume (volume solution + volume ammonium acetate) of 95% ethanol.
3. Place on ice for 15-30 min. (Leaving at 4°C overnight increases the yield somewhat, especially if the DNA concentration is low.)
4. Spin for 15-30 min in a microfuge at room temperature.
5. Pour off the supernatant. Add 1 ml of cold 70% ethanol. Spin for 10 min in a microfuge.
6. Pour off the supernatant. Invert the microfuge tube on a clean Kimwipe to blot off any excess ethanol.
7. Place in a vacuum dessicator briefly to dry off any residual ethanol.
8. Resuspend the pellet in a small volume of TE.

C. DROP DIALYSIS OF DNA

Sometimes DNA preparations contain substances that inhibit restriction enzymes or ligase. This is a quick, simple procedure to remove inhibitors such as SDS or high salt from a DNA preparation.

1. Pour about 10 ml TE buffer into a sterile petrie dish.
2. Remove a 25 mm type VS Millipore filter with forceps. Float on the surface of the TE with the shiny side up. Allow the filter to wet completely (about 2 min).
3. With a pipetman, carefully place 20-100 µl of DNA in the center of the filter.
4. Cover the petrie dish and place it somewhere it will not get bumped.
5. Leave at room temperature for 1-2 hrs.
6. Carefully remove the DNA with a pipetman.

Reference

Silhavy, T., M. Berman, and L. Enquist. 1984. *Experiments with Gene Fusions*, p. 182. Cold Spring Harbor Laboratory, NY.

D. TOOTHPICK PLASMID SCREEN

Sometimes it is necessary to screen a large number of colonies for the desired clone. This is a "quick and dirty" method that is useful for such initial mass screenings.

1. Pick a plasmid containing colony from a selective plate with a toothpick. (This procedure works best with lots of cells so pick a big colony).
2. Place in a microfuge tube containing 75 µl toothpick lysis solution.
3. Shake the toothpick to suspend the cells in the lysis buffer. Leave at room temperature about 5 min.
4. Add an equal volume of phenol:chloroform:isoamyl alcohol (25:24:1). Mix by inverting. Spin in a microfuge for 3 min.
5. Remove the upper aqueous phase (avoid the white interphase). Add 5 µl of Blue II.

6. Load 25 μl on a 0.8% agarose gel. (The aqueous phase tends to be very viscous due to the high molecular weight DNA. This can make it difficult to load into submerged wells. Running the DNA into the gel before fully submerging the wells eliminates this problem.)

Toothpick lysis solution:
6 ml   ddH₂O
3 ml   10% SDS
1 ml   1 M Tris HCl pH 8
120 μl   10 N NaOH

Appendix 6. Restriction Enzymes

A. Cleavage sites of some useful restriction enzymes

Enzyme	Reaction[1] buffer	Optimal temperature	Recognition[2] sequence	Compatible cohesive[3] ends	Heat[4] inactivated
AccI	low	37°C	GT↓$^{AT}_{CT}$AC	ClaI, HpaII, Taq, NarI	No
AvaI	med	37°C	G↓PyCGpuG	SalI, XhoI, XmaI	No
BamHI	med	37°C	G↓GATCC	BclI, Bg1II, MboI, Sau3A, XhoII	No
BclI	med	60°C	T↓GATCA	BamHI, Bg1II, MboI, Sau3A, XhoII	—
Bg1I	med	37°C	GCCNNNN↓NGGC		Yes
Bg1II	med	37°C	A↓GATCT	BamHI, BclI, MboI, Sau3A, XhoII	No
BstEII	high	60°C	G↓GTNACC		
ClaI	med	37°C	AT↓CGAT	AccI, HpaII, TaqI, NarI	Yes
DraI		37°C	TTT↓AAA	blunt	No
EcoRI	high	37°C	G↓AATTC		Yes
EcoRI*		37°C	↓AATT	EcoRI	—
EcoRV	high	37°C	GAT↓ATC	blunt	Yes
HaeIII	med	37°C	GG↓CC	blunt	No
HincII	med	37°C	GTPy↓PuAC	blunt	Yes
HindIII	med	37-55°C	A↓AGCTT		No
HinfI	med	37°C	G↓ANTC		Yes
HpaI		37°C	GTT↓AAC	blunt	No
HpaII	low	37°C	C↓CGG	AccI, ClaI, TaqI, NarI	Yes
KpnI	low	37°C	GGTAC↓C		No
MluI	med	37°C	A↓CGCGT		No
NarI		37°C	GG↓CGCC	AccI, HpaII, ClaI, TaqI	Yes
PstI	med	21-37°C	TGCA↓G	NsiI	No
PvuII	med	37°C	CAG↓CTG	blunt	No
RsaI	med	37°C	GT↓AC	blunt	Yes
SalI	high	37°C	G↓TCGAC	AvaI, XhoI	Yes
Sau3A	med	37°C	↓GATC	BamHI, Bc1II, Bg1I, Bg1II, MboI	Yes
SmaI		30°C	CCC↓GGG	blunt	Yes
TaqI	med	65°C	T↓CGA	AccI, ClaI, HpaII	No
XbaI	high	37°C	T↓CTAGA		Yes
XhoI	high	37°C	C↓TCGAG	AvaI, SalI	Yes
SacI			GAGCT↓C		
SphI			GCATG↓C		

[1] The compositions of low, medium, and high salt buffers is shown in Appendix 6B. Most restriction enzymes are reasonably active in one of these three buffers. However, when one of these buffers is not indicated special buffers are required. Many companies supply reaction buffers with restriction enzymes. Since these buffers are usually carefully optimized for each restriction enzyme, use the buffer supplied when available.

[2] Blunt-ended fragments can be ligated to any other blunt-ended fragments. Enzymes that cleave degenerate sequences give rise to a population of DNA fragments with several different termini. Only some of the possible combinations of these termini can be ligated.

[3] Compatible cohesive ends are termini produced by two different enzymes that can be ligated together. In some cases the resulting hybrid site cannot be digested by either of the two enzymes.

[4] Inactivation by heating to 65°C for 20 min.

**B. RESTRICTION
   ENZYME BUFFERS**

Often specific buffers that have been optimized for specific restriction enzymes are provided with the restriction enzyme from the supplier. However, if appropriate buffers are not supplied, one of the following three buffers will work for most restriction enzymes as indicated in Appendix 6A.

Buffer	Final concn. of 1x buffer	To prepare 0.5 ml of 10x buffer	
Low Salt	10 mM TrisHCl	50 µl	1 M TrisHCl pH 7.4
	10 mM MgCl$_2$	50 µl	1 M MgCl$_2$
	1 mM DTT	5 µl	1 M DTT
	100 µg / ml BSA	5 µl	BSA (10 mg / ml)
		390 µl	sterile ddH$_2$0
Medium Salt	50 mM NaCl	50 µl	5 M NaCl
	10 mM TrisHCl	50 µl	1 M TrisHCl pH 7.4
	10 mM MgCl$_2$	50 µl	1 M MgCl$_2$
	1 mM DTT	5 µl	1 M DTT
	100 µg / ml BSA	5 µl	BSA (10 mg / ml)
		340 µl	sterile ddH$_2$0
High Salt	100 mM NaCl	100 µl	5 M NaCl
	50 mM TrisHCl	250 µl	1 M TrisHCl pH 7.4
	10 mM MgCl$_2$	50 µl	1 M MgCl$_2$
	1 mM DTT	5 µl	1 M DTT
	100 µg / ml BSA	5 µl	BSA (10 mg / ml)
		90 µl	sterile ddH$_2$0

All the solutions should be sterile. Store the 10x buffers at -20°C.

To convert low salt buffer to medium salt buffer, add 1 µl 0.5 M NaCl per 10 µl reaction.

To convert low salt buffer to high salt buffer, add 2 µl 0.5 M NaCl + 1 µl 1 M TrisHCl (pH 7.4) per 10 µl reaction.

To convert medium salt buffer to high salt buffer, add 1 µl 0.5 M NaCl + 1 µl 1 M TrisHCl (pH 7.4) per 10 µl reaction.

Appendix 7. Electrophoresis of DNA

**SEPARATION
OF DNA FRAGMENTS**

DNA fragments have a constant charge/length ratio due to the net negative charge of the phosphate backbone. Therefore, DNA migrates toward the (+) electrode. The rate of migration during electrophoresis mainly depends on three parameters.

(1) *Pore size of the gel and length of the DNA.* During electrophoresis DNA molecules seem to "snake" through the pores in the gel "head first". As the pore size decreases (i.e. the agarose or acrylamide concentration increases) it is harder for longer DNA molecules to orient properly to snake through the pores: smaller DNA fragments snake through the pores easier and hence migrate faster. Thus, the rate of migration of linear double stranded DNA is inversely proportional to the $\log_{10}$ of its molecular weight. By using gels with different concentrations of agarose or acrylamide, a wide range of DNA fragment sizes can be separated.

(2) *Conformation of the DNA.* The shape of the DNA molecule also affects its ability to snake through the pores in the gel. In general features that make the DNA less flexible or less compact slow the migration in a gel. The rate of migration of different forms of plasmid DNA is usually: supercoiled > linear > nicked circles. In addition, secondary structure and bends in linear DNA may affect the rate of migration.

(3) *Applied voltage.* The strength of an electric field depends on the voltage and the length of the gel. At low voltages the rate of migration of linear DNA is directly proportional to the applied voltage. However, as the voltage gradient increases, the rate of migration of large DNA fragments increases relative to smaller fragments, decreasing the effective separation of different size DNA fragments.

**CURRENT, VOLTAGE,
AND POWER**

The relationship between current (I), voltage (V), resistance (R), and power (P) during electrophoresis is described by the following two equations:

$$I = V/R$$

$$P = I^2R = I\,V$$

where current is measured in milliamps, voltage in volts, and power in watts. During electrophoresis either the current, voltage, or power is held constant. Usually the resistance of agarose and acrylamide gels increases during a run. When run at constant current, the voltage increases to compensate for the increase in resistance. This results in an increase in power produced by the system, causing the gel to heat up. If the gel heats up too much the bands become distorted, so the initial current must be low enough so that the gel does not heat up excessively. In contrast, when run at constant voltage the current decreases with time to compensate for the increased resistance. Thus the gel will not heat up, but the rate of migration of the DNA decreases as the current decreases. DNA sequencing gels are often run at constant power to maintain a constant high temperature to prevent formation of secondary structure in the single stranded DNA.

AGAROSE

Agarose is an uncharged polysaccharide purified from agar. [Almost all agarose contains some anionic impurities, mainly sulfate and pyruvate, which cause electro-endosmosis (EEO) (for more explanation of EEO see The Agarose Monograph, FMC Corporation). Low EEO agarose is best for DNA gels.] Agarose melts when heated to 100°C and resolidifies when cooled below about 50°C. When it solidifies agarose forms a matrix. The size of the pores in the gel matrix can be varied by using different concentrations of agarose: the higher the concentration of agarose, the smaller the pore size. Some useful agarose concentrations for separating DNA fragments are shown in the following table.

Agarose concentration	Range of linear dsDNA fragments resolved
0.5%	2.0 - 30 kb
0.8%	0.6 - 18 kb
1.5%	0.2 - 4 kb

ACRYLAMIDE

Polyacrylamide gels are produced by polymerization of acrylamide into linear chains and cross-linking the acrylamide chains with bis-acrylamide (N,N'-methylene-bis-acrylamide). Polymerization is initiated by adding ammonium persulfate and the reaction is accelerated by TEMED (N,N,N',N'-tetramethylethylenediamine) which catalyzes the formation of free radicals from ammonium persulfate. Oxygen inhibits the polymerization because oxygen radicals formed can interact with acrylamide and terminate chain elongation. To avoid this problem, acrylamide solutions are usually "degassed" under a vacuum to remove oxygen before pouring the gel. A convenient set-up for degassing acrylamide solutions is shown below.

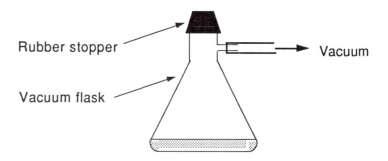

Rubber stopper

Vacuum

Vacuum flask

In addition, acrylamide gels are poured between two glass plates to exclude oxygen. The resulting meniscus which can cause the gels to run unevenly can be eliminated by overlaying the acrylamide solution with isobutanol or water. The concentration of acrylamide and the ratio of acrylamide to bis-acrylamide determine the pore size of the gel. Some useful acrylamide concentrations for separating DNA fragments are shown in the following table.

Acrylamide concentration	Range of linear dsDNA fragments resolved
5%	80 - 500 bp
12%	40 - 200 bp
20%	10 - 100 bp

References

Allington, R., J. Nelson, and C. Aron. 1975. Continuous constant power for optimal electrophoresis. *ISCO Technical Bulletin*.

Freifelder, D. 1982. *Physical Biochemistry*, Second Edition. W.H. Freeman and Co., San Francisco, CA.

Longo, M., and J. Hartley. 1986. Comparison of electrophoretic migration of linear and supercoiled molecules. *BRL Focus 8*:3-4.

Introduction to electrophoretic theory. 1988. *In* Hoefer Scientific Instruments Catalog, pp. 96-100.

A. AGAROSE GEL ELECTROPHORESIS

1. Dilute 10x TBE stock to 1x for electrophoresis buffer.
2. Add the appropriate amount of agarose to 1x TBE in a Pyrex flask. (For example, use 0.8 g agarose in 100 ml 1x TBE for a 0.8% agarose gel.)
3. Dissolve by boiling in a microwave oven.
   (CAUTION — Solutions heated in the microwave oven can become super-heated causing them to boil over when moved. ALWAYS WEAR PROTECTIVE GLOVES WHEN REMOVING FLASKS FROM THE MICROWAVE OVEN)
   Swirl the solution and check it to see if the agarose is dissolved. If not, boil the solution again until the agarose is completely dissolved.
4. Allow the hot agarose to cool to about 55°C before pouring the gel. Tape the sides of a gel tray and insert the comb. Gently pour the agarose into the gel tray. Make sure there are no bubbles.
5. Allow the gel to solidify (usually 20-30 min).
6. Remove the tape and place the gel in the electrophoresis set up. Add enough 1x TBE to submerge the gel 1-2 mm. Remove the comb.
7. Carefully load the samples into the wells with a pipetman. Include Lambda *Hind*III size standards (see Appendix 7C).
8. Attach the leads to the power supply with the samples nearest the negative (black) electrode. Run the gel at 20-50 mAmps constant current. The DNA will migrate toward the positive electrode. CAUTION — HIGH VOLTAGE.
9. When the gel has run a sufficient distance, turn off the power and unplug the leads. Carefully remove the gel and place in a baking dish.
10. Stain the gel with about 0.5 µg/ml ethidium bromide for 30-60 min. WEAR GLOVES — ETHIDIUM BROMIDE IS A MUTAGEN.
11. Photograph the stained gel on a UV transilluminator (Appendix 7D).

NOTE. If the gel stains too long it may develop a high background which obscures the DNA bands. If the background fluorescence is too high, the gel can be destained for about 30 min in dH$_2$O.

B. NONDENATURING POLYACRYLAMIDE GEL ELECTROPHORESIS (PAGE)

1. Assemble gel plates with spacers and clamps. Seal the outside edges with melted 1% agarose.
2. Prepare the polyacrylamide solution by mixing the following solutions in a vacuum flask:

Solution	Final Acrylamide/bis Concentration		
	4%	8%	12%
38% Acrylamide / 2% bis	10 ml	20 ml	30 ml
ddH$_2$O	80 ml	70 ml	60 ml
10x TBE	10.0 ml	10 ml	10 ml

Volumes are for a 16 cm x 20 cm x 0.8 mm gel

3. Cover the top of the flask with a stopper, connect to a vacuum, and degas by gently swirling until the solution no longer bubbles.
4. Add: 600 µl 10% Ammonium persulfate
           30 µl TEMED
5. Swirl gently then immediately pour into the gel plates. Save a small amount of acrylamide in a Pasteur pipet to check the polymerization.
6. Insert the comb then place a clamp over it.
7. Allow to polymerize about 1 hr.
8. Remove the comb and the bottom spacer.
9. Place the gel in an electrophoresis set up. Fill the chambers with 1x TBE. Flush the air out of the bottom slot with a Pasteur pipet bent into a V at the tip. Thoroughly rinse out the wells.
10. Slowly load the sample with a pipetman.
11. Electrophorese at 200 volts until the bromphenol blue reaches the bottom of the gel (about 3 hrs). **CAUTION — HIGH VOLTAGE.**
12. Turn off the power supply and unplug the leads. Remove the gel and lay it on a flat surface. Separate the plates by gently prying them apart with a spatula.
13. With the gel attached to one plate, stain it in a tray with 0.5 µg/ml ethidium bromide for 30-60 min. **WEAR GLOVES.**
14. Photograph on a transilluminator as described in Appendix 7D.

Reagents

Acrylamide/bis-acrylamide stock (38:2)
CAUTION - ACRYLAMIDE IS TOXIC. DO NOT MOUTH-PIPET. AVOID
      CONTACT WITH SKIN.
   Dissolve 76 g Ultra-pure acrylamide and 4 g bis-acrylamide in about 150 ml
      ddH$_2$0.
   Bring to 200 ml with ddH$_2$0.
   Filter through Whatman #1 paper.
   Store in a brown bottle at 4°C.

Ammonium persulfate (100 mg/ml)
   Dissolve 1 g ammonium persulfate in 9 ml ddH$_2$0.
   Aliquot and store at -20°C.

C. USEFUL DNA MOLECULAR WEIGHT STANDARDS

For agarose gels use Lambda *Hind*III (0.2 µg/µl) standards
Use 1 µl Lambda *Hind*III + 8 µl TE + 1 µl Blue II per lane

Fragment	Size
1	23.13 kb
2	9.42 kb
3	6.56 kb
4	4.36 kb
5	2.32 kb
6	2.03 kb
7	0.56 kb
8	0.13 kb This band may not be visible on agarose gels.

For acrylamide gels use pBR322 digested with *Hinf*I for standards
Use 1 µl pBR322 *Hinf*I + 8 µl TE + 1 µl Blue II per lane

Fragment	Size
1	1632 bp
2	517 bp
3	506 bp
4	396 bp
5	344 bp
6	298 bp
7	221 bp
8	220 bp These two bands usually appear as a doublet.
9	154 bp
10	75 bp

D. PHOTOGRAPHING ETHIDIUM BROMIDE STAINED GELS

Ethidium bromide intercalates between the stacked bases of DNA and RNA. When excited by UV light between 254 nm (short wave) and 366 nm (long wave) it emits fluorescent light at 590 nm. The DNA-ethidium bromide complex produces about 50 times more fluorescence than free ethidium bromide. Ethidium bromide can be used to detect both double stranded and single stranded nucleic acids, but the sensitivity is much greater for double stranded DNA. When ethidium bromide stained gels are photographed on a UV transilluminator a UV filter is required to screen out the background UV from the transilluminator. The following procedure describes how to photograph a ethidium bromide stained gel using a Polaroid camera set-up.

CAUTION: WEAR GLOVES. ETHIDIUM BROMIDE IS A MUTAGEN. WEAR UV PROTECTIVE GLASSES. UV LIGHT FROM THE TRANSILLUMI-NATOR CAN SERIOUSLY INJURE YOUR EYES WITHOUT EYE PROTEC-TION.

1. Place a ruler along one edge of the gel. Focus the camera with the lights on.
2. Move camera back over lens. Set the f-stop to 5.6 and place the yellow filter over the lens. Turn the room lights off and turn on the transilluminator. Usually a 1 sec exposure works well for Polaroid Type 52 film. Pull out the white tab and smoothly pull out picture tab. Wait 20 seconds, then peel the backing off your picture.
3. Measure the distance each of the Lambda *Hind*III fragments migrated relative to the wells. Plot the log (molecular weight in kilobases) vs the distance the DNA fragment moved. The curve should be linear except for very large fragments.
4. Determine the size of the other DNA fragments by measuring the distance of the bands relative to the wells and comparing to the standard curve.

(A) Migration of øX-174 DNA standards on a 6% polyacrylamide gel

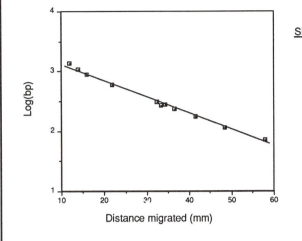

Size (bp)	Log(bp)
1353	3.13
1078	3.03
872	2.94
603	2.78
310	2.49
281	2.45
271	2.43
194	2.28
112	2.05
72	1.86

(B) Migration of øX-174 DNA fragments on a 1.5% agarose gel

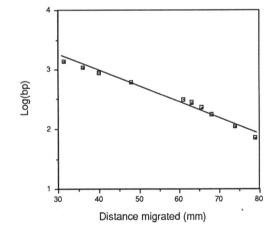

Size (bp)	Log(bp)
1353	3.13
1078	3.03
872	2.94
603	2.78
310	2.49
281	2.45
271	2.43
194	2.28
112	2.05
72	1.86

(C) Migration of Lambda HindIII standards on a 0.8% agarose gel

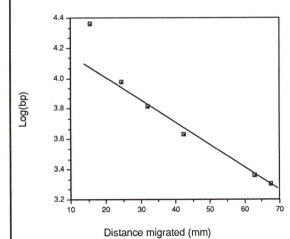

Size (kb)	Log(bp)
23.0	4.36
9.4	3.97
6.5	3.81
4.3	3.63
2.3	3.36
2.0	3.30

173

Appendix 8. Large Scale Isolation of M13 phage and RF

A. PREPARATION OF HIGH TITER M13 PHAGE STOCKS

1. Plaque the M13 on EM383 (see Experiment 10D).
2. Core a plaque with a Pasteur pipet and blow the plug into 5 ml LB.
3. Grow at 37°C for 12-15 hrs with vigorous shaking.
4. Centrifuge for 5 min at 10,000 rpm in a SS34 rotor to remove the cells. The supernatant contains the phage stock for large scale growth of template. (Do NOT add chloroform! M13 is very sensitive to organic solvents.)

B. LARGE SCALE INFECTION AND GROWTH OF M13 PHAGE AND RF

1. Inoculate a 2 liter flask containing 500 ml of LB with 5 ml of an overnight culture of EM383.
2. Grow the cells at 37°C to approximately 40 Klett units.
3. Add 0.5 ml of the high titer M13 phage stock per 500 ml culture and shake vigorously for 4 to 6 hrs at 37°C.
4. Centrifuge at 5,000 rpm for 10 min in a GSA rotor.

The supernatant contains the single stranded phage and the cell pellet contains the double stranded RF. The RF can be purified using the large scale plasmid isolation protocol (begin with step #4).

C. PURIFICATION OF ssDNA FROM M13 PHAGE

Purification of intact phage:
1. Add 14.5 g NaCl and 25 g of PEG 8000 per 500 ml supernatant.
2. Chill on ice for at least 1 hr.
3. Centrifuge at 10,000 rpm for 20 min at 4°C in a SS32 rotor.
4. Resuspend the pellet in 7 ml of TE buffer.
5. Add 5.2 g CsCl.
6. Bring to a final volume of 13 ml with TE buffer. Fill a small quick seal tube and seal the top. (You must have an even number of tubes to balance the rotor. If necessary, prepare a balance tube.)
7. Place the tubes in a Ti70 rotor and cover with spacers. Centrifuge at 40,000 rpm for 48 hr at 20-25°C.
8. The phage will form a clear band in the middle of the gradient. It is easist to see if you hold the centrifuge tube near a light. Puncture the top of the tube with an 18 guage needle. Puncture the side of the tube with an 18 guage needle just below the phage band and gently withdraw the phage band into a syringe.
9. Dialyze the phage overnight against ten volumes TE buffer. Change the buffer three times.

Extraction of ssDNA from M13:
1. Dilute a small aliquot of the phage to an $OD_{280} = 1.5$. Based on this dilution, dilute the rest of the phage to $OD_{280} = 15$ (i.e., 10-fold more concentrated than the diluted aliquot).
2. Phenol extract once with 50% phenol/chloroform saturated with TE + 0.2% sarkosyl. Avoid the interphase!
3. Phenol extract once with phenol saturated with TE + 0.2% sarkosyl.
4. Phenol extract once with phenol saturated with TE.
5. Dialyze overnight against ten volumes of TE buffer. Change the buffer three times.
6. Store at 4°C.

Index

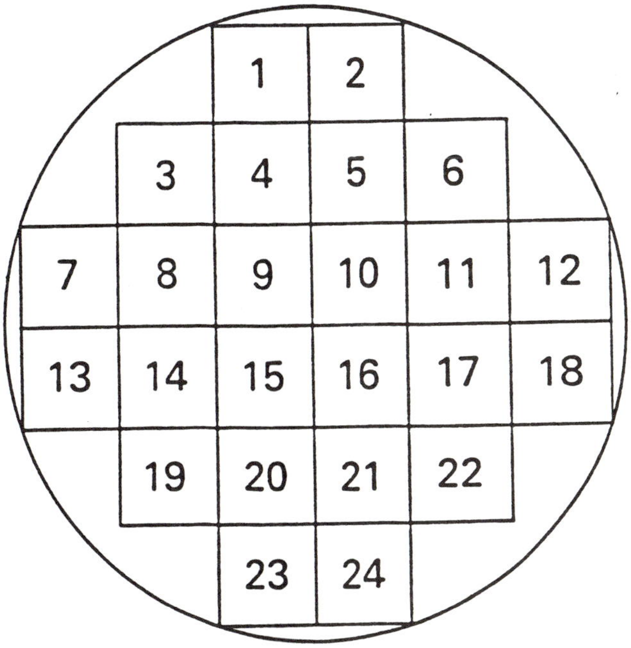

Top left circle

```
        1   2
    3   4   5   6
7   8   9  10  11  12
13  14  15  16  17  18
   19  20  21  22
      23  24
```

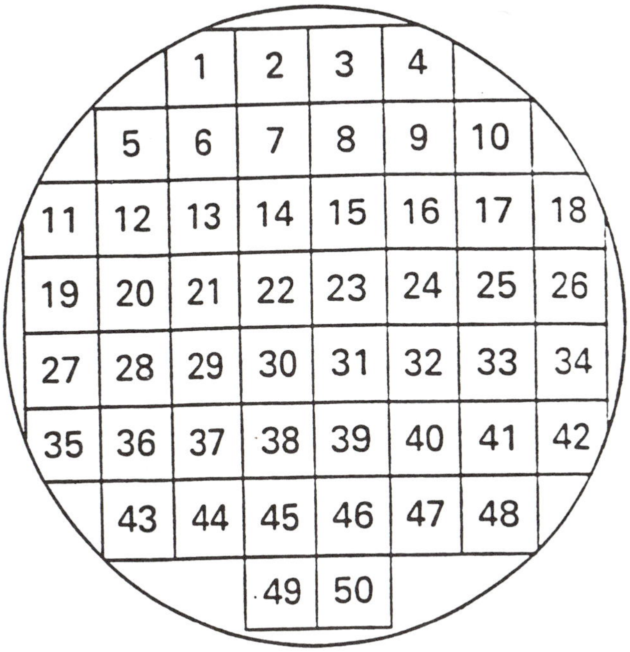

Top right circle

```
    1   2   3   4
  5   6   7   8   9  10
11  12  13  14  15  16  17  18
19  20  21  22  23  24  25  26
27  28  29  30  31  32  33  34
35  36  37  38  39  40  41  42
   43  44  45  46  47  48
      49  50
```

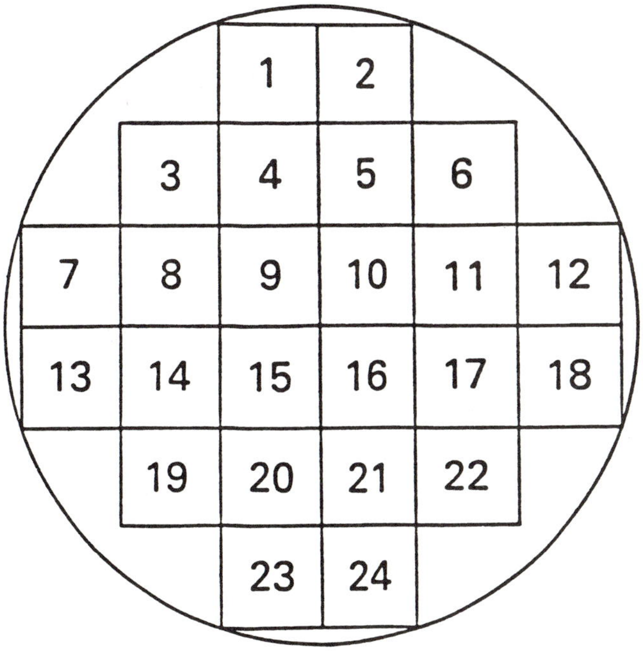

Bottom left circle

```
        1   2
    3   4   5   6
7   8   9  10  11  12
13  14  15  16  17  18
   19  20  21  22
      23  24
```

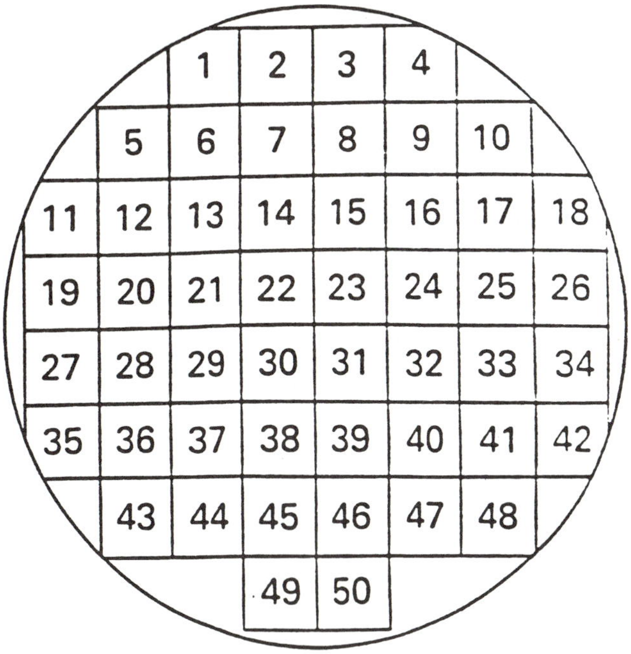

Bottom right circle

```
    1   2   3   4
  5   6   7   8   9  10
11  12  13  14  15  16  17  18
19  20  21  22  23  24  25  26
27  28  29  30  31  32  33  34
35  36  37  38  39  40  41  42
   43  44  45  46  47  48
      49  50
```